# Springer Theses

Recognizing Outstanding Ph.D. Research

For further volumes:
http://www.springer.com/series/8790

## Aims and Scope

The series "Springer Theses" brings together a selection of the very best Ph.D. theses from around the world and across the physical sciences. Nominated and endorsed by two recognized specialists, each published volume has been selected for its scientific excellence and the high impact of its contents for the pertinent field of research. For greater accessibility to non-specialists, the published versions include an extended introduction, as well as a foreword by the student's supervisor explaining the special relevance of the work for the field. As a whole, the series will provide a valuable resource both for newcomers to the research fields described, and for other scientists seeking detailed background information on special questions. Finally, it provides an accredited documentation of the valuable contributions made by today's younger generation of scientists.

## Theses are accepted into the series by invited nomination only and must fulfill all of the following criteria

- They must be written in good English.
- The topic should fall within the confines of Chemistry, Physics, Earth Sciences, Engineering and related interdisciplinary fields such as Materials, Nanoscience, Chemical Engineering, Complex Systems and Biophysics.
- The work reported in the thesis must represent a significant scientific advance.
- If the thesis includes previously published material, permission to reproduce this must be gained from the respective copyright holder.
- They must have been examined and passed during the 12 months prior to nomination.
- Each thesis should include a foreword by the supervisor outlining the significance of its content.
- The theses should have a clearly defined structure including an introduction accessible to scientists not expert in that particular field.

James B. Glattfelder

# Decoding Complexity

## Uncovering Patterns in Economic Networks

Doctoral Thesis accepted by
Swiss Federal Institute of Technology (ETH)
Zurich, Switzerland

Springer

*Author*
Dr. James B. Glattfelder
Chair of Systems Design
Swiss Federal Institute
of Technology (ETH)
Zurich
Switzerland

*Supervisor*
Prof. Frank Schweitzer
Head of Chair of Systems Design
Swiss Federal Institute
of Technology (ETH)
Zurich
Switzerland

ISSN 2190-5053 ISSN 2190-5061 (electronic)
ISBN 978-3-642-33423-8 ISBN 978-3-642-33424-5 (eBook)
DOI 10.1007/978-3-642-33424-5
Springer Heidelberg New York Dordrecht London

Library of Congress Control Number: 2012947397



Printed on acid-free paper

Springer is part of Springer Science+Business Media (www.springer.com)

*To Ladina*

# Supervisor's Foreword

While globalization seems to have been *the* buzzword since the late twentieth century, scientific insight into the causes and effects, and the benefits and drawbacks of globalization still needs to be fully developed. For an understanding of how globalization works, massive data analysis of economic agents worldwide is an inevitable step. What does the structure of global economic interactions look like? Does it resemble structures found on smaller economic scales, or are there emerging properties on the global level? Do our findings conform to the established economic theories, or must these theories be refined and further developed to cope with these findings? Can we derive models to predict the dynamics in such global networks?

The thesis of James Glattfelder addresses the heart of these scientific questions. Based on the theory of complex networks, it derives a novel approach to investigate large-scale economic networks based on massive analysis of ownership data. It further links these findings to relevant economic questions, such as the measurement of control. For this, James proposes novel ways to quantify economic concepts like "value", "control", "power", and provides computational algorithms to extract these from weighted, directed network structures hidden in large-scale datasets.

A particular highlight of the thesis is the unfolding of a global ownership network of transnational companies (TNC), including their shareholders and their subsidiaries. This analysis identified a rather small, but fully connected core of the network which contains less than 1 % of all nodes, but almost 20 % of the operating revenue of all TNC. This "core" is mostly dominated by financial intermediaries based in the US and UK. The existence of such a structure is an important clue toward understanding systemic risk in economic systems, in particular if seen together with the very strong concentration of control that James has identified. The "emerging picture" reveals the role of a "super-entity" in the global network of TNC, a finding that goes beyond our current knowledge about globalization. The insights obtained in this work even challenge established economic theories, because they are hardly predicted or explained by these.

The content of the thesis is highly original and makes a fundamental contribution to our knowledge of global economic networks. Moreover, it is one of the rare cases where scientific results have made it to the front page of major newspapers and magazines around the world. It thereby fuelled the discussion on inequality, the role of financial institutions, and the control of economic power. I am pleased that the publication of this thesis makes such important findings available to a larger audience.

Zurich, July 2012 Prof. Dr. Dr. Frank Schweitzer

# Acknowledgments

When I first embarked on the journey of writing this thesis, originally called "Ownership Networks and Corporate Control: Mapping Economic Power in a Globalized World", I would never have imagined being so lucky. I was able to profit greatly while interacting with a kind group of dedicated people. The continued international media coverage of our work revealed to me that the topic we chose to study is indeed of greater interest and relevance to our society. Surprisingly, there has been little advancement in this direction to understand the interconnected nature of our economic systems and decode the apparent complexity. Finally, I would like to thank the people at *Springer* for accepting the nomination of this work and publishing it in their *Springer Theses* series.

Zurich, June 2012 James B. Glattfelder

Within a few months of writing these words, my 38th birthday will be coming up. As this is an unusually old age for finishing ones thesis, I would like to mention all the people who contributed to the meandering path leading to this milestone. In chronological order: I would like to thank my mother and late father, Caroline and Tazi Glattfelder, who, from an early age on, fostered my budding interest in science. My special gratitude goes out to Peter Henley-Smith from Forest School, London, who coincidentally and spontaneously offered to introduce me to chemistry and physics at the age of 15 and awakened my dormant passion for science.

I thank my employer since 2002, Richard B. Olsen of Olsen Ltd., Zurich, who, from the very beginning on, was highly supportive of my endeavor to complete a Ph.D. and arranged for me to continue working at the company on a part-time basis. I am very grateful to Frank Schweitzer, who was willing to embark on the adventure of advising an older than average and only part-time employed Ph.D. student. I also thank him for his scientific guidance and support.

I am particularly indebted to Stefano Battiston for his continuous and invaluable advice and dedicated tutoring, allowing me to benefit greatly. Moreover, I would like to thank the lively team at the Chair of Systems Design for their friendship and scientific interaction. In particular, Stefania Vitali for a successful collaboration. Finally, I would like to thank Guido Caldarelli for kindly offering to be an ideal co-examiner.

Furthermore, and in general, I am grateful to all my friends who helped and supported me during this undertaking. Last but not least, I am sincerely thankful to Ladina Plattner for being a constant source of joy.

Zurich, June 2010 James B. Glattfelder

# Contents

# Acronyms

| | |
|---|---|
| BFS | Breadth-first search (search algorithm) |
| B-index | Banzhaf index |
| CC | (Weakly) Connected component |
| CDF | Cumulative distribution function |
| COMM | Community (e.g., COMM.1 refers to the largest community) |
| DRG | Directed random graph |
| FS | Financial sector |
| GPAM | Generalized preferential attachment model |
| H-index | Herfindahl index |
| IN | In-section of a bow-tie |
| LCC | Largest CC |
| LM | Linear model (for estimating control from ownership see also RM and TM) |
| NACE | (Industry standard classification system) |
| OCC | Other connected components (everything outside the LCC) |
| OECD | Organization for Economic Co-operation and Development |
| OR | Operating revenue |
| OUT | Out-section of a bow-tie |
| PC | Participated company |
| PDF | Probability density function |
| RM | Relative model (for estimating control from ownership see also LM and TM) |
| SCC | Strongly connected component (in the main text, this is synonymous with the core of the bow-tie in the LCC) |
| SH | Shareholder (economic actors holding shares in TNCs) |
| SS-index | Shapley–Shubik index |
| TM | Threshold model (for estimating control from ownership see also LM and RM) |
| TNC | Transnational corporation (OECD definition) |
| TPH | Top power-holder (list of TNCs and SHs that together hold 80 % of the network control) |

| | |
|---|---|
| T&T | Tubes and tendrils (sections in a bow-tie that either connect IN and OUT, are outgoing from IN, or in-going to OUT, respectively) |
| WWW | World Wide Web |

In addition, countries are abbreviated by their two letter ISO 3166-1 alpha-2 codes, e.g., CH, JP, US. etc.

# Chapter 1
# Introduction

*We spend billions of dollars trying to understand the origins of the universe, while we still don't understand the conditions for a stable society, a functioning economy, or peace.*

(D. Helbing quoted in Cho 2009)

It sounds paradoxical, but today it appears that we understand more about the universe than our society. We have created systems, that have outgrown our capacity to genuinely understand and control them. In order to try and provide a contribution to the understanding of socio-economic systems, this thesis aims at answering the following questions:

**How is economic control distributed globally**?
**Who are the key economic actors holding the largest fraction of control**?
**To what degree are the top economic actors interconnected with each other**?

The answers to these questions lie at the interface between the realms of social and systems science—namely, economics and the study of complex networks. These two notions are embodied in the main object of study, introduced in Sect. 1.1.3: the ownership network.

In order to understand all the involved issues and concepts, this chapter introduces the related prerequisites. Before the economics topics from the fields of corporate governance and corporate finance are discussed in Sect. 1.2, some general questions need to be addressed:

**What are real-world complex networks and how do they relate to the study of complex systems?**

The long answer to these questions deals not only with the philosophy of science, discussed in Appendix A.1, but also relates to the two realms of reality that have been successfully understood using formal models: fundamental and complex processes, presented in Sects. A.2.2 and A.2.3 (in general, Appendix A reflects the author's views on these issues and is an attempt to illustrate and illuminate these intangible notions). The short answer is given in the next section.

J. B. Glattfelder, *Decoding Complexity*, Springer Theses,
DOI: 10.1007/978-3-642-33424-5_1, © Springer-Verlag Berlin Heidelberg 2013

## 1.1 Complex Networks

A *complex system* is usually understood as being comprised of many interacting or interconnected parts. A characteristic feature of such systems is that the whole often exhibits properties not obvious from the properties of the individual parts. This is called *emergence*. In other words, a key issue is how the macro behavior emerges from the interactions of the system's elements at the micro level.

The empirical analysis of real-world complex systems has revealed unsuspected regularities, such as scaling laws, which are robust across many domains (Mantegna and Stanley 1995; West et al. 1997; Amaral et al. 1998; Albert et al. 1999; Pastor-Satorras et al. 2001; Newman et al. 2002; Garlaschelli et al. 2003; Newman 2005; Glattfelder et al. 2011). For more details see Appendix C. This has suggested that universal or at least generic mechanisms are at work in the structure-formation and evolution of many such systems. Tools and concepts from statistical physics have been crucial for the achievement of these findings (Dorogovtsev and Mendes 2003; Caldarelli 2007).

Complex systems are usually very reluctant to be cast into closed-form analytical expressions. This means that it is generally hard to derive mathematical quantities describing the properties and dynamics of the system under study.

Complex systems can, however, be easily described by their structure of interactions. The complexity of the agents that make up the system can be ignored. Therefore any real-world complex system finds its natural formal representation in a graph: the nodes represent the agents and the links their interactions. In essence, the properties of the complex system can be understood by mapping it onto a complex network and then studying the network regularities. This is very much in the spirit of the paradigm, that simple rules give rise to complex behavior. For more details on these issues, see Appendix A.2.3.

Today, complex networks are ubiquitous[1]: phenomena in

- the physical world, e.g.,
  - computer-related systems (Albert et al. 1999; Barabási et al. 2000; Tadić 2001; Pastor-Satorras et al. 2001; Capocci et al. 2006),
  - various transportation structures (Banavar et al. 1999; Guimera et al. 2005; Kühnert et al. 2006),
  - power grids (Albert et al. 2004),
  - spontaneous synchronization of systems (Gómez-Gardenes et al. 2007),
- biological systems, e.g.,
  - neural patterns (Ripley 2008),

[1] For instance, (Strogatz 2001; Albert and Barabási 2002; Dorogovtsev and Mendes 2002, 2003; Newman 2003; Newman et al. 2006; Caldarelli 2007).

  - epidemiology (Meyers et al. 2005),
  - food chains (Garlaschelli et al. 2003; McKane and Drossel 2005),
  - gene regulation (Bennett et al. 2008),
  - spontaneous synchronization in biological systems (Gonze et al. 2005),

- social[2] and economic realms, e.g.,

  - diffusion of innovation (Schilling and Phelps 2007; König et al. 2009),
  - trust-based interactions (Walter et al. 2008),
  - various collaborations (Newman 2001a,b),
  - social affiliation (Brown et al. 2007),
  - trade relations (Serrano and Boguñá 2003; Garlaschelli and Loffredo 2004a,b; Reichardt and White 2007; Fagiolo et al. 2008, 2009),
  - shared board directors (Strogatz 2001; Battiston and Catanzaro 2004),
  - similarity of products (Hidalgo et al. 2007),
  - credit relations (Boss et al. 2004; Iori et al. 2008),
  - price correlation (Bonanno et al. 2003; Onnela et al. 2003),
  - ownership networks (Glattfelder and Battiston 2009; Vitali et al. 2011).

are best understood if characterized as networks. The explosion of this field of research was and is driven by the increasing availability of huge amounts of data, pouring in from neurobiology, genomics, ecology, finance and the World-Wide Web, etc., in combination with the access to massive computing power and vast storage facilities.

### *1.1.1 Three Levels of Network Analysis*

In the last years, in order to offer useful insights into more detailed research questions, several studies have started taking into account the specific meaning of the nodes and links in the various domains the real-world networks pertain to. The study of real-world complex networks can be performed at three levels of analysis. Level 1 is the purely topological approach (best epitomized by a binary adjacency matrix (see Appendix B.2) where links simply exists or don't). Allowing the links to carry information, i.e., have directions and weights, defines Level 2 (Newman 2004; Barrat et al. 2004; Barthelemy et al. 2004; Onnela et al. 2005; Ahnert et al. 2007).

At the highest level of detail, the nodes themselves are assigned a degree of freedom, in the guise of non-topological state variables that shape the topology of the network (Garlaschelli and Loffredo 2004c; Garlaschelli et al. 2005; De Masi et al. 2006). These variables are sometimes also called fitness. See Fig. 1.1 for a visualization of the 3-level approach to complex networks. However, Level 3 type analysis have

[2] A general reference is Vega-Redondo (2007).

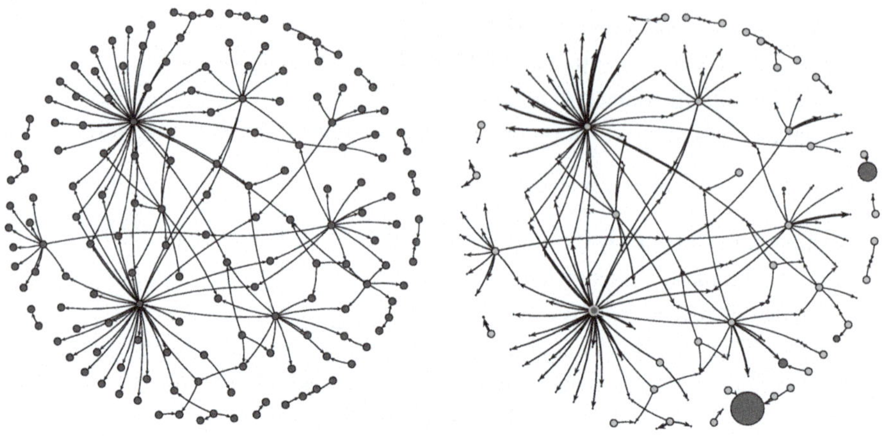

**Fig. 1.1** Visualization examples of the same underlying network: (*left*) a directed layout; (*right*) the full-fledged 3-level layout, where the thickness of the links represents their weight and the nodes are scaled by some non-topological state variable; the colors can reflect some additional attribute of the nodes; the graph layouts are based on Geipel (2007)

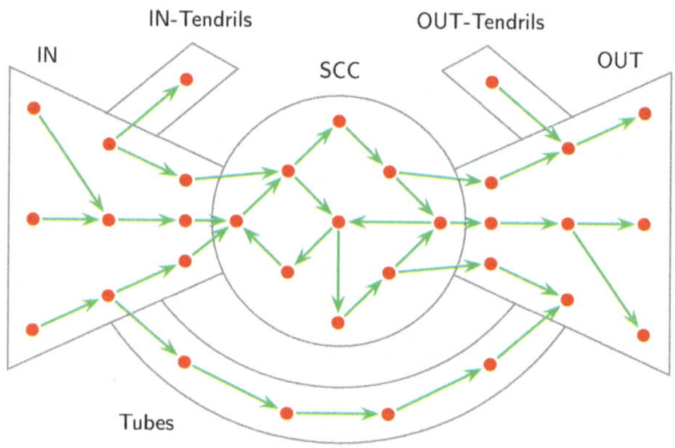


**Fig. 1.2** Schematic illustration of a bow-tie topology: the central area is the strongly connected component (SCC), where there is a path from each node to every other node, and the *left* (IN) and *right* (OUT) sections contain the incoming and outgoing nodes, respectively; the set of nodes directly linking the IN and OUT are called tubes; the IN can have outgoing tendrils, and the OUT incoming ones; the tubes and tendrils are referred to as T&T

still today not become state-of-the-art (Caldarelli et al. 2002; Servedio et al. 2004; Garlaschelli and Loffredo 2004b). One reason for this is that considering all three levels of detail does not guarantee *per se* that new insights can be gained. It is also essential that the standard measures utilized in the analysis of complex networks are appropriately adapted to the specific nature of the network under investigation.

In this thesis, we show how employing all three levels when analyzing ownership networks yields new insights which would otherwise remain unobserved.

Finally, a further regularity can be discovered in real-world networks if they are analyzed at Level 2. This concerns their topological structure.[3] When considering the direction of links, so-called *bow-tie topologies* can be identified. This is a core-periphery structure, with an incoming and outgoing segment. Its core is formed by nodes that are all reachable from each other following chains of directed links. This is called a strongly connected component (SCC).[4] Note that a network can have multiple SCCs. Each of these cores defines a bow-tie structure, with their incoming (IN) and outgoing (OUT) nodes emanate from them. Figure 1.2 illustrates such a topology.

### 1.1.2 Economic Networks

As mentioned, an extraordinary wide range of natural phenomena can be understood as complex networks. From physics, biology, earth and planetary sciences, finance, computer science and demography to the social sciences. However, to this day economic networks are still acutely underrepresented in this field of study. This issue will be discussed further in Sect. 1.2.3.

To quote the summary given in Schweitzer et al. (2009), appropriately called "Economic Networks: The New Challenges":

> *In summary, we anticipate a challenging research agenda in economic networks, built upon a methodology that strives to capture the rich process resulting from the interplay between agents' behavior and the dynamic interactions among them. To be effective, however, empirical studies providing insights into economic networks from massive data analysis, theory encompassing the appropriate description of economic agents and their interactions, and a systemic perspective bestowing a new understanding of global effects as coming from varying network interactions are needed. We predict that such studies will create a more unified field of economic networks that advances our understanding and leads to further insight. We are still far from a satisfactory identification and integration of the many components, but the recent advances outlined suggest a promising start.*

This thesis responds to these challenges by providing an analysis of economic control in various national networks, next to the global network. Chapter 2 develops a new theoretical framework to measure the flow of control in ownership networks. The 3-level empirical analysis based on this methodology are described in Chaps. 3 and 4. The scope of the datasets covers millions of economic entities in over hundred countries. Finally, Chap. 5 describes a network-formation model giving rise to the

---

[3] See Appendix B.3 for an introduction of the notion.

[4] A list of acronyms can be found in front matter of this book.

observed empirical patterns. This offers new insights into possible micro-foundations for the interactions of economic agents.

Our analysis sheds new light on competition in global markets and systemic financial risk. The results are likely to be of interest to a broad audience and have attracted the attention of both the media and economic advisors. In fact, the publications related to Chaps. 3 and 4 were widely covered in the news (see Appendix F) and we have also been contacted by a governmental agency.

### *1.1.3 Ownership Networks*

The specific economic networks studied in this thesis are ownership networks. A detailed introduction to the related economic topics and definitions is given in Sect. 1.2. For the moment it suffices to understand that a percentage of ownership expresses the fraction that a shareholder owns in the firm it has shares in. Shareholders can either be entities who cannot be owned themselves (i.e., natural persons, families, cooperative societies, registered associations, foundations, public authorities or other legal entities) or national or transnational corporations (see Sect. 1.2.1).

Ownership networks are very interesting objects to study, for various reasons. Next to representing an interface between the fields of economics and complex networks, the structure of (international) economic power, or corporate control,[5] is reflected in the network of ownership ties of companies. Ownership networks have been used to measure the impact of globalization or institutional interventions,[6] and they are related to issues of corporate social responsibility and sustainability (Wheeler et al. 2003). They can be understood to exhibit:

> *[...] the historical bargains struck by labor, the state and holders of capital regarding who gets to own and control the economic assets.*
>
> (Kogut and Walker 2001, p. 317)

The study of ownership networks has implications for individual firms:

> *A network of ownership ties represents a unique opportunity to examine how an economy-wide structure of relations affects individual firm diversification events.*
>
> (Kogut and Walker 1999, p. 6)

> *The relationship of capital to the firm is also shaped by the structure of interfirm networks, which influences firm behavior through access to critical resources and information.*
>
> (Aguilera and Jackson 2003, p. 454)

Finally, they are relevant for policymakers:

[5] This notion will be introduced in Sect. 1.2.1.

[6] This is discussed in Sect. 1.2.3.

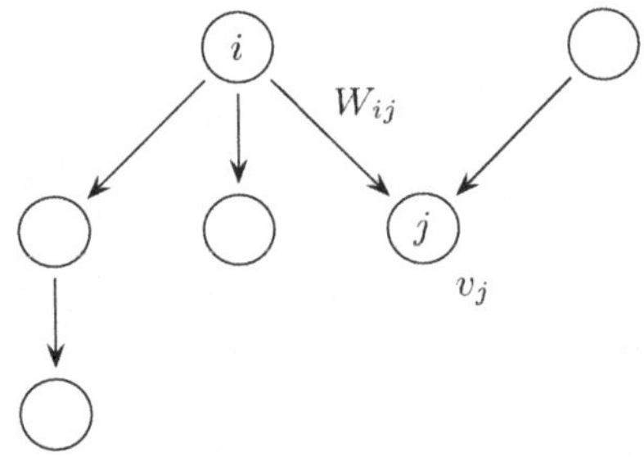


**Fig. 1.3** Schematic illustration of a stylized ownership network: nodes represent shareholders and firms, the directed weighted links $W_{ij}$ denote the percentage of ownership; the nodes representing companies carry a non-topological state variable $v_j$, acting as a proxy for their value or size

> *The analysis of networks of business enterprises has grown to be one of the leading perspectives in the study of business policy, organizational behavior, and public economic policy.*
> (Corrado and Zollo 2006, p. 319)

To obtain a network from ownership data, the percentage of ownership shareholder $i$ has in company $j$ is encoded in the adjacency matrix $W$ by the entry $W_{ij}$. Figure 1.3 shows a simple illustration of an ownership relation. Note that because the firms can also appear as shareholders, the network does not display a bipartite structure.

Furthermore, the quantity $v_j$ acts as a proxy for the intrinsic or underlying value of the firm $j$. Common choices are market capitalization, total assets or the operating revenue. While the market capitalization, being defined as the number of outstanding shares[7] times the firm's market price, is a good proxy for the generic value or size of a firm, market prices are only available for listed companies, i.e., firms with shares traded on a stock exchange. Taking total assets as a proxy for the size of firms results in a heavy bias in favor of banks. Operating revenue, defined as the net sales revenue accruing from the primary business operations of a firm, is a measure which is widely available. In Chap. 3, $v_j$ is taken to be the market capitalization of listed companies. In Chap. 4 we employ the operating revenue.

It holds by definition that the sum of shares in a company must add up to 100 %

$$\sum_{i=1}^{n} W_{ij} = 1; \quad j = 1, \ldots, n. \tag{1.1}$$

In graph-theoretic terms, the adjacency matrix $W$ is column stochastic. Note however, that sometimes the equality in Eq. (1.1) is not found for the empirical data. The sum can be smaller than 100 % due to unreported shareholdings. Such missing ownership data is nearly always due to their percentage values being very small and therefore negligible. Hence a straight-forward procedure is to normalize the ownership percentages.

An example of a natural measure to derive from these quantities is shareholder $i$'s *portfolio value*:

---

[7] See Sect. 1.2.1.

$$p_i := \sum_{j \in \Gamma(i)} W_{ij} v_j, \tag{1.2}$$

where $\Gamma(i)$ is the set of indices of the neighbors of $i$. Or in matrix notation $p = Wv$.

The question of what control means in this context and how it can be measured in ownership networks is answered in Chap. 2.

## 1.2 Elements from Economics

As discussed in the previous sections, it is essential to embed the complex network under study in the context given by its real-world domain. This not only enhances the understanding of the processes forming them but also gives true meaning to the 3-level network analysis.

The issues relating to ownership are discussed in the economics literature from the fields of corporate governance and corporate control. All the concepts revolve around the main entity being analyzed: the corporation. In the following section these themes will be introduced.

Finally, various empirical studies from economics have dealt with the detection and analysis of small ownership patterns, which are discussed in Sect. 1.2.2.

### *1.2.1 The Corporation: Ownership and Control*

A *corporation* is defined as an institution that is a separate legal entity having its own privileges, similar to a natural person. There are many different forms of corporations, most of which are used to conduct business. Corporations have limited liability, can sue, borrow or lend money, buy and sell shares, takeover and merge with other corporations. It must also pay taxes. In a nutshell, *corporate finance* analyzes how the corporations make long-term financial decisions and investigates their financing patterns (Brealey and Myers 1996).

Figure 1.4 shows a simplified illustration of the structure of a corporation. From its business activities the corporation strives to make profits. The money required for new investments can come from different sources. The corporation can finance itself by reinvesting profits. There are also two modes for acquiring external funding. Firstly, debt is sold in the form of bonds to investors or financial institutions. The financial sector also plays an important role in lending money to corporations. Secondly, shares of stock can be issued.

The *stock* represents the original capital paid into or invested in the business by its founders and serves as a security. Hence the stock is also referred to as *equity*

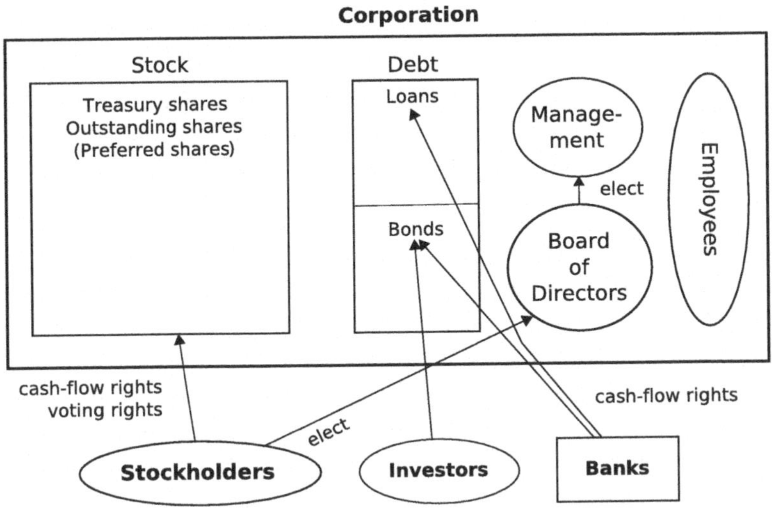


**Fig. 1.4** A simple illustration of a corporation, see main text for a discussion

*securities*, or equity for short. Different kinds of shares can be issued by the corporation. *Outstanding shares* are common shares that have been authorized, issued, and purchased by investors. They have voting rights associated with them and so-called cash-flow rights giving claims on the corporation's assets, earnings and dividends. Outstanding shares represent ownership in the corporation by the person or institution that holds the shares. These shares should be distinguished from *treasury shares*, which is common stock held by the corporation itself. Treasury shares may have come from a repurchase of shares by the firm from shareholders or they may have never been issued to the public. These shares do not pay dividends and have no assigned voting rights. Finally, preferred shares give investors a greater claim on the corporations' assets but have no voting rights. The so-called free float of a company is defined as those shares that are readily available for trading.

The entities owning shares in the stock of a company are called stockholders, or synonymously *shareholders*. As mentioned, the shareholders gain cash-flow rights and voting rights over the corporation in exchange for the invested money allowing an equity stake to be held in the corporation. This is represented by an ownership relation $W_{ij}$: the fraction of outstanding shares shareholder $i$ holds in corporation $j$. Furthermore, the shareholders collectively owning a company have complete control over its strategic business decisions and financial strategies. This control is exerted either by voting at shareholder's meetings or by appointing the board of directors, which in turn elects the senior management. So-called proxy voting is a procedure to delegate the authority to another member of a voting body to vote in their absence.

The term *corporate control* refers to the power to make investment and financing decisions.

The field of *corporate governance* deals with the questions of how control and ownership are separated, the possible accountability conflicts and opposing interests between shareholders and managers (also called the principle agent, or agency problem). Corporate governance also analyzes the role played by the board of directors and any action taken by the shareholder to influence corporate decisions, (See also Stiglitz 1985; Eisenhardt 1989; Brealey and Myers 1996; Shleifer and Vishny 1998; Aguilera and Jackson 2003). In summary, the financial architecture of a corporation comprises elements from corporate governance, corporate control and the corporation's relationship with financial institutions.

Note that shareholders do not only act as individuals but can collaborate in shareholding coalitions that give rise to so-called voting blocks (Nenova 2003). The theory of political voting games in cooperative game theory has been applied to the problem of shareholder voting. This will be discussed in Sect. 2.8.

On a final note, in this thesis, the terms firm, company and corporation will be used interchangeably.

## *1.2.2 Ownership and Control Patterns*

In most industrialized countries, firms are connected with each other by many ownership links which form complex patterns. In essence, the financial architecture mentioned at the end of the last section is extended by the corporation's interconnectedness. What is known about these organizational patterns of ownership and the resulting structure of control? In the following, a short summary introducing the relevant studies is given.

### Corporate Ownership Around the World

In their classic text, Berle and Means (1932), discuss the separation of ownership from control for the US. The authors find that a large body of equity holders exercise hardly any control. This is contrasted with a small group of controlling managers. The general conclusion that was drawn, was that in Anglo-Saxon countries ownership and hence control is dispersed amongst a large number of outside investors. In other words, these countries have the highest occurrence of so-called *widely held firms*. This statement, that the control of corporations is dispersed amongst many shareholders, invokes the intuition that there exists a multitude of owners that only hold a small amount of shares in a few companies.[8] In effect, the authors sparked the discussion

[8] However, in anticipation of our findings presented in Chap. 3, and in contrast to such intuition, the empirical network analysis of Anglo-Saxon countries reveals an entirely different picture: although,

on the separation of ownership from control and the paradigm of widely held firms would define the direction of research for the next decades.

A subsequent global empirical study, decades later, uncovered a more diverse picture: globally, dispersed ownership is actually quite rare and indeed usually only observed for Anglo-Saxon countries. Concentrated ownership (e.g., family control) is much more prevalent in most of the 27 countries analyzed (La Porta et al. 1999). Of all the conceivable national determinants (legal settings, law enforcement, corruption, tax rules, institutional settings, market size, maturity of the banking sector, etc.) the distinguishing feature is reported to be mainly the consequence of legal protection: ownership concentration is a result of poor legal protection of minority shareholders. For European countries, an older study also uncovered concentrated control structures (Franks and Mayer 1995). Barca and Becht (2001) also present an analysis of control in European countries in a more recent book. They criticize some of the results of La Porta et al. (1999) on methodological grounds. Finally, Windolf (2002) analyzes corporate networks of firms connected by managers (so-called *interlocking directorates*) in Europe and the United States. The author also investigates ownership relations and analyzes the structural similarities between the interlock networks and the capital networks.

### Cross-Shareholdings

Complex structures of ownership can themselves be used as vehicles to separate ownership from control. A simple shareholding structure used to separate ownership from control is a so-called pyramid. It is comprised of an ultimate owner sitting on top of the ownership structure exerting control on firms located at lower levels down the chain. This is reminiscent of the ownership structure seen in Fig. 2.1. See also Almeida and Wolfenzon (2006).

More complex ownership patterns are *cross-shareholdings*. These are said to occur when corporations own shares in each other. Generally, cross-shareholdings are sub-networks where companies own each other directly or indirectly through chains of ownership relations. In other words, they are loose coalitions of firms. Figure 1.5 shows some examples of cross-shareholdings of different degree. Note that cross-shareholdings correspond to SCCs in graph theory, recalling the core of the bow-tie topology of Fig. 1.2.

Cross-shareholdings are common in European, Asian and South American countries. They are sometimes referred to as conglomerates or business groups. Prominent examples are the *keiretsu* in Japan, the *chaebol* in Korea and *grupos economicos* in South America (Granovetter 1995). As an example, the *keiretsu* is a network of companies, usually organized around a major bank, reflecting long-term business

from a local perspective, firms are indeed widely held, from a bird's-eye point-of-view one can identify the existence of a small elite of shareholders continually reappears as the controlling entity of all the stocks, without ever having been previously detected or reported on.

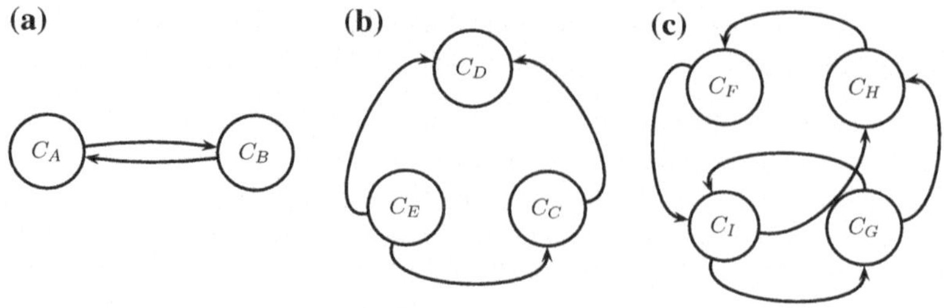


**Fig. 1.5** Examples of cross-shareholdings of different degrees of complexity: **a** mutual cross-shareholding, **b** possible cross-shareholding with three firms, **c** cross-shareholding of higher degree

relationships. It is a system of corporate governance, where the power is split between the main bank, the largest companies and the group as a whole. There are many proposed theories explaining their existence. Among the most important are: the anti-takeover theory (preventing hostile takeovers), the externality theory (positive spillover of information), and the successive monopoly theory (stipulating informal agreements on prices) (Flath 1996).

There is a vast literature focussed on cross-ownership relations: (Goto 1982; Flath 1992; Feenstra 1998; La Porta et al. 1999; O'Brien and Salop 1999; Feenstra et al. 1999; Claessens and Djankov 2000; Dore 2002; Chapelle 2005; Gilo et al. 2006; Almeida et al. 2007; Trivieri 2007). Note also, that to what extent the economic activities of a country are in the control of one or more groups of few actors, is also a recurrent question in history. The answer has important implications in terms of market competition (O'Brien and Salop 1999, 2001; Gilo et al. 2006; Trivieri 2007), systemic risk (Battiston et al. 2007; Wagner 2009; Haldane 2009; Stiglitz 2010), and for political power (Windolf 2002). In Sect. 6.2 these topics are re-discussed again, in the light of the empirical results presented in this thesis.

It is an interesting observation, that all Anglo-Saxon countries share a unique "type" of capitalism, the so-called Atlantic, stock market or arm's length capitalism (Rajan and Zingales 1998; Dore 2002). Furthermore, it is striking that business groups are assumed to be absent in these countries (Granovetter 1995). The reasons for this can be seen in various historical developments, as explained in the following for the US.

## The Special Case of Ownership and Control in the United States

In the late 19th century, a populist view in the US saw the so-called Money Trust, an alliance of a few New York banks, as the center of corporate power (Brandeis 1914). This bank-centered system was perceived as unhealthy competition and lead to the adoption of a series of antitrust regulations by the mid-1930s. Still today, in many developed countries, the financial sector is often described by the popular press as a kind of puppet master behind the scenes.

In parallel, American households greatly expanded their equity ownership. In other words, the public was participating in the stock market. This widespread dispersion of stock was then documented by Berle and Means (1932) at the beginning of the 1930s, who, as a consequence, saw the management being in control. This was understood as a lamentable situation because of the perceived unaccountability of managers (agency problem). This rise of so-called managerialism was unchallenged until the 1960s. Then, slowly the paradigm changed. The idea that dispersed ownership, embodied in the notion of shareholder value, was seen as a favorable and efficient setup. In contrast, concentrated ownership was perceived as a threat, indicating that minority shareholders were not well protected (La Porta et al. 1999; Davis and Useem 2002). This led to a change in thinking about corporate governance in the US, motivated by the idea that a democratic corporate governance, together with diluted shareholding, can prevent excessive concentration of control. As a result, there has been a recent wave of liberalization of the financial markets. For more details see Davis (2008).

The 1960s and 1970s saw a wave of conglomerate mergers. This reflected the strategy of growth through acquiring firms in unrelated lines of business and structuring them as a collection of separate business units. However, in the 1980s and 1990s there was a period of "deconglomeration" dismantling these business groups. The reason for this was seen in the inherent instability of American-style conglomerates (Davis et al. 1994).

This story will be continued in Sect. 6.1.5 when we address the question of the relevance of our empirical findings of Chaps. 3 and 4 …

### *1.2.3 Ownership Networks Revisited*

We now come back to the topic of ownership networks discussed in Sect. 1.1.3 and view them in an economics context.

#### Existing Work

To summarize the insights from economics so far, the large body of corporate governance can be grouped into three main categories:

(i) identifying the seat of power (who is really in control of a corporation and how control is dispersed or concentrated);
(ii) empirical investigations of how the patterns of control vary across countries and what determines them;

(iii) analyzing the frequently observed complex ownership patterns (e.g., business groups and pyramids) and how they act as vehicles to separate ownership from control.

It should be noted that these previous studies did not build on the idea that ownership and control define a vast network of dependencies. Instead, they selected samples of selected important companies (for instance in terms of market capitalization) and looked only at their local web of interconnections. The aim was usually to identify the ultimate owner of these small networks of firms. As an example, La Porta et al. (1998) studies the ten largest corporations in 49 countries, La Porta et al. (1999) looks at the 20 largest public companies in 27 countries, Claessens and Djankov (2000) analyzes 2980 companies in nine East Asian countries, Faccio and Lang (2002) traces the ultimate ownership and control of 5232 European corporations, and Chapelle (2005) utilizes a set of 800 Belgian firms.

It is a remarkable fact that the investigation of the financial architecture of corporations in national or global economies taken *as a network* is just at the beginning. Notable exceptions are the following three studies, two being from the field of complex networks. The pioneering work in Kogut and Walker (2001) looks at the German ownership network in 1993 and analyzes the merger and acquisition events from 1994 to 1997. The dataset is comprised of 550 firms and 685 shareholders. The authors investigate the topological features, namely the small world property (see Appendix B.3) of the network and identify communities. They find that the core structure of important economic actors is extremely robust and resilient, defying the powerful force of globalization. The study conclude that "power is self-preserving" and "embedded in small worlds" (Kogut and Walker 1999).

Another prominent empirical analysis of ownership networks is Corrado and Zollo (2006). This work studies the impact of institutional interventions on the Italian ownership network. Two snapshots of the network are compared: 1990 (454 shareholders, 212 firms, and 817 ownership relations) and 2000 (553 owners, 207 companies, and 751 links). Although the effect of the interventions is witnessed by the increased fragmentation of the network at the macro level, again, there is a high stability in the structure of the backbone of the network comprised of the most important economic actors. The authors conclude:

> *The network of cross-ownerships might have generalizable features of a small world, and these features might be sufficiently resilient to institutional change. If this is the case, then the small world of business ownership networks might need to be accepted as a 'natural' element of the industrial texture in a given country. There is no point in trying to fight gravity.*
>
> (Corrado and Zollo 2006, p. 349)

Finally, Windolf (2002) presents an economics-based analysis of corporate networks in five European countries and the US. Although the book provides some rudimentary ownership network analysis (largest component, degree, density, etc.), the emphasis lies on the networks given by interlocking directorates, respectively the structural similarities between the two. Furthermore, it determines the owners of the largest corporations. As an example, in Germany the 650 largest firms and 821 shareholders are analyzed. For the US, these numbers are 250 firms and 5925 shareholders.

These studies, however, have two major drawbacks. Firstly, they only considered the network at Level 1 (i.e., by employing a binary adjacency matrix), ignoring the weighted and directed nature of ownership networks. For Kogut and Walker (2001); Corrado and Zollo (2006), this constraint comes from the fact that the authors analyzed the small-world phenomena, which is per definition a Level 1 network quantity. In Latora and Marchiori (2003) the authors generalize the notion of small-world networks to the directed and weighted case. However, this formalism has never been applied to economic networks. The only studies analyzing ownership networks at higher levels are finance related and deal with market investments. In Stark and Vedres (2005) weights are employed and the evolution of the network of foreign direct investments is analyzed for Hungary. Examples of a full Level 3 network analysis are found in Battiston et al. (2005); Garlaschelli et al. (2005), focussing on the scaling-law behavior of market and inter-regional investments.

Secondly, the scope of the analyzed networks is very small with a couple of hundred nodes and links. In contrast, the empirical analysis that will be presented in Chaps. 3 and 4 are based on

- 48 country networks, totalling 131018 nodes and 545896 links;
- the global network surrounding all transnational corporations (TNCs), with 600508 nodes and 1006987 links.

## 1.3 On the Horizon

Not only has the empirical investigation of the financial architecture of corporations in national or global economies never been performed extensively from a complex networks point of view, in addition, from a purely theoretical perspective, economics does not offer any models that predict the structure of national and global corporate control. This is a surprising and prominent gap in the literature. With this thesis, we start to fill in some of the details.

Although it is intuitive that every large corporation has a pyramid of subsidiaries located below itself in the network and a number of shareholders above, it is not clear how much the network surrounding one corporation is expected to interact with the networks surrounding other corporations. Three alternative hypotheses can be formulated. Corporations may remain isolated, cluster in separated coalitions, or form a giant connected component, possibly with a core-periphery structure. As all of these structures have different implications for the distribution of control, both a topological analysis and a joint investigation of the distribution of control need to be carried out in order to uncover the true organization of the market.

With the insight gained in the last sections, an additional question can be added to the ones already raised at the beginning of this chapter. In summary and to conclude, the analysis of ownership networks can give answers to the following:

(i) What is the distribution of control?
(ii) Are the control structures fragmented or integrated?
(iii) Are there aggregate organizational structures comprised of important corporations to be discerned?
(iv) Who are the key economic actors?
(v) What is the role played by the financial sector?

Or, in other words:

**What is the map of corporate control?**

All these questions can be posed either at a *national* or the *global* level. The empirical analysis presented in this thesis cover both points of view.

To quote Andy Haldane, executive director of financial stability at the Bank of England in London, with respect to our study (Vitali et al. 2011):

> *This gave us a tantalising glimpse of a brave new world for finance. I want to try and map that world. If we do so, I believe fragilities in the global financial system which have stalked the globe for four years (and counting) could be significantly reduced.*
>
> *Now imagine the light this financial map might shine. It would allow regulators to issue the equivalent of weather-warnings—storms brewing over Lehman Brothers, credit default swaps and Greece. It would enable advice to be issued—keep a safe distance from Bear Stearns, sub-prime mortgages and Icelandic banks. And it would enable "what-if?" simulations to be run—if UK bank Northern Rock is the first domino, what will be the next?*
>
> *At present, global finance does not have the technology to ask these questions, much less answer them. The prize is a big, but attainable, one. It would not entirely lift the fog from finance, but it could provide us with a navigation system better able to avoid the next crash landing.*

The whole article is given in Sect. F.3.2 of the appendix.

All of this signifies a paradigm shift[9] approaching a complexity-network-view of economics, moving towards analyzing interactions, where ideas like interconnection, co-dependence and collective dynamics enter the stage. Indeed, slowly this new paradigm is being embraced by scholars and policy makers. For instance, the EU Commissions' scientific project for Forecasting Financial Crises.[10] Also ambitious, long-term and highly-funded big-data programs like *futurICT*—an EU *FET*

[9] See Appendix A.1.3.

[10] Called *FOC*: http://www.focproject.net/.

*Flagship Initiative*[11]—the UN's *GlobalPulse*[12] or the US' *Big Data R&D Initiative*[13] are currently being set up. These efforts aim at crafting new technologies and innovations building on a complex systems point of view and trying to improve our understanding of the organizing principles of socio-economic systems, highlight potential weaknesses and looming threats. Ultimately being able to give concrete advice and formulate effective policies. However, these programs still mostly exist at a preliminary level, where the project is still being formulated or the data collected.

This thesis aims at uncovering patterns in ownership networks and is one of the first examples of this new kind of endeavor in economics. As mentioned, the findings raise important issues like the concentration of power, systemic risk and market competition. But in a first step, the question of how corporate control can be formalized within the context of complex networks must be answered. This is done in the next chapter, developing the new methodology employed in this thesis.

Skipping to the discussions of our work, in Chap. 6 we give a short summary and elucidate the real-world relevance of our novel findings and discuss their implications.

## References

R.V. Aguilera, G. Jackson, The cross-national diversity of corporate governance: dimensions and determinants. Acad. Manag. Rev. **28**(3), 447–465 (2003)

S.E. Ahnert, D. Garlaschelli, T.M.A. Fink, G. Caldarelli, Ensemble approach to the analysis of weighted networks. Phys. Rev. E **76**(1), 16101 (2007)

R. Albert, A.L. Barabási, Statistical mechanics of complex networks. Rev. Mod. Phys. **74**(1), 47–97 (2002)

R. Albert, H. Jeong, A.L. Barabási, Internet: diameter of the World-Wide Web. Nature **401**(6749), 130–131 (1999)

R. Albert, I. Albert, G.L. Nakarado, Structural vulnerability of the North American power grid. Phys. Rev. E **69**(2), 25103 (2004)

H. Almeida, S.Y. Park, M. Subrahmanyam, D. Wolfenzon, Beyond Cash Flow and Voting Rights: Valuation and Performance of Firms in Complex Ownership Structures. Working paper, New York University Stern School of Business, 2007

H.V. Almeida, D. Wolfenzon, A theory of pyramidal ownership and family business groups. J. Finance **61**(6), 2637–2680 (2006)

L.A.N. Amaral, S.V. Buldyrev, S. Havlin, M.A. Salinger, H.E. Stanley, Power law scaling for a system of interacting units with complex internal structure. Phys. Rev. Lett. **80**(7), 1385–1388 (1998)

J.R. Banavar, A. Maritan, A. Rinaldo, Size and form in efficient transportation networks. Nature **399**(6732), 130–131 (1999)

A.L. Barabási, R. Albert, H. Jeong, Scale-free characteristics of random networks: the topology of the World-Wide Web. Physica A **281**(1–4), 69–77 (2000)

[11] See in detail (Farmer et al. 2012) or http://www.futurict.eu/ and http://www.fet11.eu/about/fet-flagships.

[12] See http://www.unglobalpulse.org/.

[13] For instance, http://www.whitehouse.gov/sites/default/files/microsites/ostp/big_data_press_release.pdf.

F. Barca, M. Becht, *The Control of Corporate Europe* (Oxford University Press, Oxford, 2001)
A. Barrat, M. Barthelemy, R. Pastor-Satorras, A. Vespignani, The architecture of complex weighted networks. Proc. Natl. Acad. Sci. **101**, 3747 (2004)
M. Barthelemy, A. Barrat, R. Pastor-Satorras, A. Vespignani, Characterization and modeling of weighted networks. Physica A **346**, 34–43 (2004)
S. Battiston, M. Catanzaro, Statistical properties of corporate board and director networks. Eur. Phys. J. B **38**(2), 345–352 (2004)
S. Battiston, D. Delli Gatti, M. Gallegati, B. Greenwald, J.E. Stiglitz, Credit chains and bankruptcy propagation in production networks. J. Econ. Dyn. Control **31**(6), 2061–2084 (2007)
S. Battiston, J.F. Rodrigues, H. Zeytinoglu, The network of inter-regional direct investment stocks across Europe. Adv. Complex Syst. **10**(1), 29–51 (2005)
M.R. Bennett, W.L. Pang, N.A. Ostroff, B.L. Baumgartner, S. Nayak, L.S. Tsimring, J. Hasty, Metabolic gene regulation in a dynamically changing environment. Nature **454**(7208), 1119–1122 (2008)
A. Berle, G. Means, *The Modern Corporation and Private Property* (Mac Millan, New York, 1932)
G. Bonanno, G. Caldarelli, F. Lillo, R.N. Mantegna, Topology of correlation-based minimal spanning trees in real and model markets. Phys. Rev. E **68**(4), 46130 (2003)
M. Boss, H. Elsinger, M. Summer, S. Thurner, Network topology of the interbank market. Quant. Finance **4**(6), 677–684 (2004)
L.D. Brandeis, *Other People's Money and How the Bankers Use It* (Oxford University Press, Oxford, 1914)
R.A. Brealey, S.C. Myers, *Principles of Corporate Finance* (McGraw-Hill, New York, 1996)
J. Brown, A.J. Broderick, N. Lee, Word of mouth communication within online communities: conceptualizing the online social network. J. Interact. Mark. **21**(3), 2 (2007)
G. Caldareli, *Scale-Free Networks: Complex Webs in Nature and Technology* (Oxford University Press, Oxford, 2007)
G. Caldarelli, A. Capocci, P. de Los Rios, M.A. Muñoz, Scale-free networks from varying vertex intrinsic fitness. Phys. Rev. Lett. **89**(22), 258702 (2002)
A. Capocci, V.D.P. Servedio, F. Colaiori, L.S. Buriol, D. Donato, S. Leonardi, G. Caldarelli, Preferential attachment in the growth of social networks: the Internet encyclopedia Wikipedia. Phys. Rev. E **74**(3), 36116 (2006)
A. Chapelle, Separation of ownership and control: Where do we stand? Corp. Ownership Control **2**(2), 91–101 (2005)
A. Cho, Ourselves and our interactions: the ultimate physics problem? Science **325**(5939), 406 (2009)
S. Claessens, S. Djankov, The separation of ownership and control in East Asian corporations. J. Financ. Econ. **58**(1–2), 81–112 (2000)
R. Corrado, M. Zollo, Small worlds evolving: governance reforms, privatizations, and ownership networks in Italy. Ind. Corp. Change **15**(2), 319–352 (2006)
G.F. Davis, A new finance capitalism? Mutual funds and ownership re-concentration in the United States. Eur. Manag. Rev. **5**(1), 11–21 (2008)
G.F. Davis, M. Useem, Top management, company directors, and corporate control, in *Handbook of Strategy and Management*, ed. by A. Pettigrew, H. Thomas, R. Whittington (Sage, London, 2002), pp. 233–259
G.F. Davis, K.A. Diekmann, C.H. Tinsley, The decline and fall of the conglomerate firm in the 1980s: the deinstitutionalization of an organizational form. Am. Sociol. Rev. **59**(4), 547–570 (1994)
G. De Masi, G. Iori, G. Caldarelli, Fitness model for the Italian interbank money market. Phys. Rev. E **74**(6), 66112 (2006)
R. Dore, Stock market capitalism vs. welfare capitalism. New Polit. Econ. **7**(1), 115–127 (2002)
S.N. Dorogovtsev, J.F.F. Mendes, Evolution of networks. Adv. Phys. **51**(4), 1079–1187 (2002)
S.N. Dorogovtsev, J.F.F. Mendes, *Evolution of Networks: From Biological Nets to the Internet and WWW* (Oxford University Press, Oxford, 2003)

K.M. Eisenhardt, Agency theory: an assessment and review. Acad. Manag. Rev. **14**(1), 57–74 (1989)
M. Faccio, L.H.P. Lang, The ultimate ownership of western European corporations. J. Financ. Econ. **65**(3), 365–395 (2002)
G. Fagiolo, J. Reyes, S. Schiavo, On the topological properties of the world trade web: a weighted network analysis. Physica A **387**, 3868 (2008)
G. Fagiolo, J. Reyes, S. Schiavo, World-trade web: topological properties, dynamics, and evolution. Phys. Rev. E **79**(3), 36115 (2009)
J.D. Farmer, M. Gallegati, C. Hommes, A. Kirman, P. Ormerod, S. Cincotti, A. Sanchez, D. Helbing, A complex systems approach to constructing better models for managing financial markets and the economy. Eur. Phys. J. (Submitted 2012)
R.C. Feenstra, Integration of trade and disintegration of production in the global economy. J. Econ. Perspect. **12**(4), 31–50 (1998)
R.C. Feenstra, T.-H. Yang, G.G. Hamilton, Business groups and product variety in trade: evidence from South Korea. Taiwan and Japan. J. Int. Econ. **48**(1), 71–100 (1999)
D. Flath, Indirect shareholding within Japan's business groups. Econ. Lett. **38**(2), 223–227 (1992)
D. Flath, The Keiretsu puzzle. J. Jpn. Int. Econ. **10**(2), 101–121 (1996)
J. Franks, C. Mayer, Ownership and control, in *Trends in Business Organization: Do Participation and Cooperation Increase Competitiveness?*, ed. by H. Siebert (University of Michigan Press, Michigan, 1995)
D. Garlaschelli, S. Battiston, M. Castri, V.D.P. Servedio, G. Caldarelli, The scale-free topology of market investments. Physica A **350**, 491–499 (2005)
D. Garlaschelli, M.I. Loffredo, Fitness-dependent topological properties of the world trade web. Phys. Rev. Lett. **93**(18), 188701 (2004a)
D. Garlaschelli, M.I. Loffredo, Structure and evolution of the world trade network. Physica A **355**, 138–144 (2004b)
D. Garlaschelli, M.L. Loffredo, Patterns of link reciprocity in directed networks. Phys. Rev. Lett. 93(1) (2004c)
D. Garlaschelli, G. Caldarelli, L. Pietronero, Universal scaling relations in food webs. Nature **423**(6936), 165–168 (2003)
M.M. Geipel, Self-organization applied to dynamic network layout. Int. J. Mod. Phys. C **18**(10), 1537–1549 (2007)
D. Gilo, Y. Moshe, Y. Spiegel, Partial cross ownership and tacit collusion. RAND J. Econ. **37**(1), 81–99 (2006)
J.B. Glattfelder, S. Battiston, Backbone of complex networks of corporations: the flow of control. Phys. Rev. E **80**(3), 36104 (2009)
J.B. Glattfelder, A. Dupuis, R.B. Olsen, Patterns in high-frequency FX data: discovery of 12 empirical scaling laws. Quant. Financ. **11**(4), 599–614 (2011)
J. Gómez-Gardenes, Y. Moreno, A. Arenas, Paths to synchronization on complex networks. Phys. Rev. Lett. **98**(3), 34101 (2007)
D. Gonze, S. Bernard, C. Waltermann, A. Kramer, H. Herzel, Spontaneous synchronization of coupled circadian oscillators. Biophys. J. **89**(1), 120–129 (2005)
A. Goto, Business groups in a market economy. Eur. Econ. Rev. **19**(1), 53–70 (1982)
M. Granovetter, Business groups in the modern economy, in *Industrial and Corporate Change*, ed. by D.J. Teece, J. Chytry (Oxford University Press, New York, 1995)
R. Guimera, S. Mossa, A. Turtschi, L.A.N. Amaral, The worldwide air transportation network: anomalous centrality, community structure, and cities' global roles. Proc. Natl. Acad. Sci. **102**(22), 7794 (2005)
A.G. Haldane, Rethinking the financial network. *Speech at Financial Student Association, Amsterdam* (2009), http://www.bankofengland.co.uk/publications/speeches/2009/speech386.pdf
C.A. Hidalgo, B. Klinger, A.L. Barabási, R. Hausmann, The product space conditions the development of nations. Science **317**(5837), 482 (2007)
G. Iori, G. De Masi, O.V. Precup, G. Gabbi, G. Caldarelli, A network analysis of the Italian overnight money market. J. Econ. Dyn. Control **32**(1), 259–278 (2008)

B. Kogut, G. Walker, The Small World of Firm Ownership in Germany: Social Capital and Structural Holes in Large Firm Acquisitions—1993–1997. Unpublished working paper, Wharton School, University of Pennsylvania, 1999

B. Kogut, G. Walker, The small world of Germany and the durability of national networks. Am. Sociol. Rev. **66**(3), 317–335 (2001)

M.D. König, S. Battiston, F. Schweitzer, Modeling Evolving Innovation Networks. Innovation Networks, ed. by A. Pyaka, A. Scharnhorst, pp. 187–268, (Springer, 2009)

C. Kühnert, D. Helbing, G.B. West, Scaling laws in urban supply networks. Physica A **363**, 96–103 (2006)

R. La Porta, F. Lopez de Silanes, A. Shleifer, R.W. Vishny, Law and finance. J. Polit. Econ. **106**(6), 1113–1155 (1998)

R. La Porta, F. Lopez de Silanes, A. Shleifer, Corporate ownership around the world. J. Finance **54**(2), 471–517 (1999)

V. Latora, M. Marchiori. Economic small-world behavior in weighted networks. Eur. Phys. J. B **32**(B), 249–263 (2003)

R.N. Mantegna, H.E. Stanley, Scaling behaviour in the dynamics of an economic index. Nature **376**(6535), 46–49 (1995)

A.J. McKane, B. Drossel, Models of food web evolution, in *Ecological Networks*, ed. by V. Wolters, J.C. Moore, (Elsevier, Amsterdam, 2005), pp. 74–88

L.A. Meyers, B. Pourbohloul, M.E.J. Newman, D.M. Skowronski, R.C. Brunham, Network theory and SARS: predicting outbreak diversity. J. Theor. Biol. **232**(1), 71–81 (2005)

T. Nenova, The value of corporate voting rights and control: a cross-country analysis. J. Financ. Econ. **68**(3), 325–351 (2003)

M.E.J. Newman, Scientific collaboration networks. I. Network construction and fundamental results. Phys. Rev. E **64**(1), 16131 (2001a)

M.E.J. Newman, Scientific collaboration networks. II. Shortest paths, weighted networks, and centrality. Phys. Rev. E **64**(1), 16132 (2001b)

M.E.J. Newman, The structure and function of complex networks. SIAM Rev. **45**(2), 167–256 (2003)

M.E.J. Newman, Analysis of weighted networks. Phys. Rev. E **70**(5), 56131 (2004)

M.E.J. Newman, Power laws, Pareto distributions and Zipf's law. Contemp. Phys. **46**(5), 323–351 (2005)

M.E.J. Newman, D.J. Watts, S.H. Strogatz, Random graph models of social networks. Proc. Natl. Acad. Sci. **99**(90001), 2566–2572 (2002)

M.E.J. Newman, A.L. Barabási, D.J. Watts, *The Structure and Dynamics of Networks* (Princeton University Press, Princeton, 2006)

D.P. O'Brien, S.C. Salop, Competitive effects of partial ownership: financial interest and corporate control. Antitrust Law J. **67**, 559–614 (1999)

D.P. O'Brien, S.C. Salop, The competitive effects of passive minority equity interests: reply. Antitrust Law J. **69**, 611 (2001)

J.-P. Onnela, A. Chakraborti, K. Kaski, J. Kertész, A. Kanto, Dynamics of market correlations: taxonomy and portfolio analysis. Phys. Rev. E **68**(5), 56110 (2003)

J.-P. Onnela, J. Saramäki, J. Kertész, K. Kaski, Intensity and coherence of motifs in weighted complex networks. Phys. Rev. E **71**(6), 065103.1–065103.4 (2005)

R. Pastor-Satorras, A. Vázquez, A. Vespignani, Dynamical and correlation properties of the Internet. Phys. Rev. Lett. **87**(25), 258701 (2001)

R.G. Rajan, L. Zingales, Which Capitalism? Lessons from the East Asian Crisis. CRSP working papers 486, Center for Research in Security Prices, Graduate School of Business, University of Chicago, 1998. http://ideas.repec.org/p/wop/chispw/486.html

J. Reichardt, D.R. White, Role models for complex networks. Eur. Phys. J. B **60**(2), 217–224 (2007)

B.D. Ripley, *Pattern Recognition and Neural Networks* (Cambridge University Press, Cambridge, 2008)

M.A. Schilling, C.C. Phelps, Interfirm collaboration networks: the impact of large-scale network structure on firm innovation. Manag. Sci. **53**(7), 1113–1126 (2007)
F. Schweitzer, G. Fagiolo, D. Sornette, F. Vega-Redondo, A. Vespignani, D.R. White, Economic networks: the new challenges. Science **325**(5939), 422 (2009)
M.Á. Serrano, M. Boguñá, Topology of the world trade web. Phys. Rev. E **68**(1), 15101 (2003)
V.D.P. Servedio, P. Buttà, G. Caldarelli, Vertex intrinsic fitness: How to produce arbitrary scale-free networks. Phys. Rev. E **70**, 056126 (2004)
A. Shleifer, R.W. Vishny, A survey of corporate governance. J. Finance **52**(2), 737–783 (1998)
D. Stark, B. Vedres, Social times of network spaces: network sequences and foreign investment in Hungary. Am. J. Sociol. **111**(5), 1367–1411 (2005)
J.E. Stiglitz, Credit markets and the control of capital. J. Money Credit Bank. **17**(2), 133–152 (1985)
J.E. Stiglitz, Risk and Global Economic Architecture: Why Full Financial Integration May Be Undesirable. NBER working paper, 2010
S.H. Strogatz, Exploring complex networks. Nature **410**(6825), 268–276 (2001)
B. Tadić, Dynamics of directed graphs: the world-wide web. Physica A **293**(1–2), 273–284 (2001)
F. Trivieri, Does cross-ownership affect competition? Evidence from the Italian banking industry. J. Int. Financ. Markets Inst. Money **17**(1), 79–101 (2007)
F. Vega-Redondo, *Complex Social Networks* (Cambridge University Press, Cambridge, 2007)
S. Vitali, J.B. Glattfelder, S. Battiston, The network of global corporate control. PloS ONE **6**(10), e25995 (2011)
W. Wagner, Diversification at financial institutions and systemic crises. J. Financ. Intermed. **19**, 373–386 (2009)
F.E. Walter, S. Battiston, F. Schweitzer, A model of a trust-based recommendation system on a social network. Auton. Agent. Multi-Agent Syst. **16**(1), 57–74 (2008)
G.B. West, J.H. Brown, B.J. Enquist, A general model for the origin of allometric scaling laws in biology. Science **276**(5309), 122 (1997)
D. Wheeler, B. Colbert, R.E. Freeman, Focusing on value: reconciling corporate social responsibility, sustainability and a stakeholder approach in a network world. J. Gen. Manag. **28**(3), 1–28 (2003)
P. Windolf, *Corporate Networks in Europe and the United States* (Oxford University Press, Oxford, 2002)

# Chapter 2
# The Main Methodology: Computing Control in Ownership Networks

*The first point is that the enormous usefulness of mathematics in the natural sciences is something bordering on the mysterious and that there is no rational explanation for it.*

*Fundamentally, we do not know why our theories work so well.*

(E. Wigner in Wigner 1960)

In this chapter the mathematical bulk of the thesis is presented. The aim of having an exhaustive account of the methods results in the extensive scope of the chapter. As networks find their mathematical embodiment in adjacency matrices (see Appendix B), most of the formalism is comprised of linear-algebraic manipulations.

The reader who is mostly interested in the application of the methods, i.e., the empirical network analysis, can directly go to Chaps. 3 and 4. In Chap. 5 a network-evolution model is presented, shedding new light on the micro rules underlying the empirical properties. All these chapters are written to be self-supporting and provide a minimal introduction to the details of the methodology given in the following. Alternatively, a brief summary is found in Sects. 2.9 and 2.11. Chapter 6 also summarizes the results before discussing the relevance and implications of our findings.

## 2.1 Introduction

Given an adjacency matrix of an ownership network, what can be said about the distribution of control? The long answer to this question encompasses the following aspects:

1. the introduction of existing measures;
2. their extension and correction;
3. a reinterpretation and unification using network-theoretic notions.

J. B. Glattfelder, *Decoding Complexity*, Springer Theses,
DOI: 10.1007/978-3-642-33424-5_2, © Springer-Verlag Berlin Heidelberg 2013

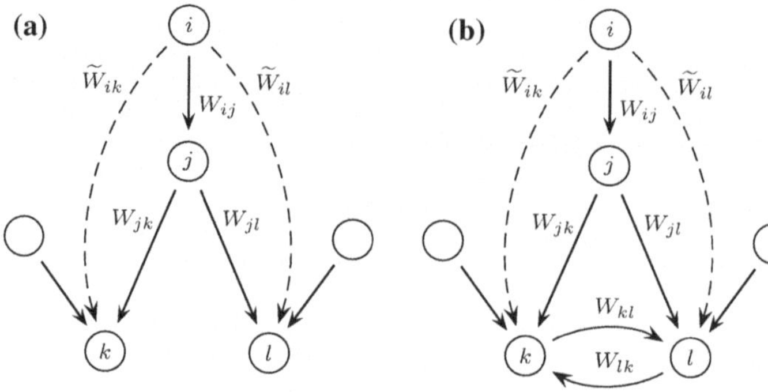


**Fig. 2.1** **a** Company $i$ has $W_{ij}$ percent of direct ownership in company $j$ and indirect ownership in companies $k$ and $l$, through $j$; the computation of indirect ownership is straightforward as long as the network has no cycles; as an example, $\widetilde{W}_{ik} = W_{ij} W_{jk}$; **b** in the presence of cycles, it is necessary to account for the arising recursive paths of indirect ownership; the model of integrated ownership does this, see Sect. 2.2.1

In a nutshell, the answer is as follows: from the knowledge of the ownership relations the control associated with a shareholder can be estimated. This quantity then needs to be adapted to account for all the indirect paths in the network.

Recall from Sect. 1.1.3 that the percentage of ownership firm $i$ has in company $j$ is given by the entry $W_{ij}$ in the adjacency matrix. The underlying value of the firms are denoted by $v_j$. In Chaps. 3 and 4, $v_j$ is taken to be the market capitalization and the operating revenue, respectively. In the next section, a first try at assessing the impact of a network structure on ownership is presented.

## 2.2 Direct and Indirect Ownership

Figure 2.1 illustrates how chains of direct ownership lead to indirect paths of ownership. In the case of tree-like topologies, e.g., Fig. 2.1a, the indirect paths are multiples of the direct links comprising them. In the presence of cycles, this trivial procedure breaks down and the methodology introduced in the following replaces it.

### *2.2.1 A First Try: Group Value and Integrated Ownership*

In Brioschi et al. (1989) and Brioschi and Paleari (1995), the authors propose a simple algebraic model of ownership structures that reflects the direct and indirect ownership relations. It is based on the input–output matrix methodology introduced to economics in Leontief (1966). The sum of all direct and indirect ownership shares a shareholder has in the equity capital of a firm is collectively called *integrated ownership*.

The authors analyzed a setting given by a single external shareholder owning shares in a cluster of firms in a business group, i.e., firms connected by cross-shareholdings. The authors derive two equations, one assigning values to the firms in the business group, and one for computing the integrated ownership shares attributed to this external owner.

### Group Value

Let $v$ be a column vector containing the intrinsic value[1] of the firms in the business group. The adjacency matrix $W^G$ describes all the links between the group of firms connected by cross-shareholdings. In accordance with Eq. (1.2) the portfolio value $p_i^G$ of firm $i$ in the group is given by

$$p_i^G = \sum_{j \in \Gamma(i)} W_{ij}^G v_j. \tag{2.1}$$

The row vector of the direct ownership ties of the external shareholder is given by $d$. This shareholder's portfolio value is given by

$$p_{\text{ext}} = dv. \tag{2.2}$$

The so-called *group value* is defined as follows

$$v^G := W^G v^G + v. \tag{2.3}$$

In other words, the group value reflects the value or importance of a firm depending on its position in the business group, the group's interconnectedness and the distribution of $v_j$. In Sect. 2.6 we will reinterpret this quantity as a network centrality measure: a firm's value depends on the neighboring firms value plus an initial value. The solution is found to be

$$v^G = (I - W^G)^{-1} v, \tag{2.4}$$

where $I$ denotes the identity matrix. The associated group value of the external shareholder can be computed directly as

$$v_{\text{ext}}^G := dv^G + v_{ext}, \tag{2.5}$$

where $v_{ext}$ is the external shareholder's value.

Putting all these values into one vector, we get

---

[1] Brioschi et al. (1989) use the value of the net assets.

$$v^{G,\text{tot}} = \begin{pmatrix} d(I - W^G)^{-1}v + v_{ext} \\ (I - W^G)^{-1}v \end{pmatrix}. \tag{2.6}$$

**Integrated Ownership**

As mentioned, integrated ownership refers to the total of direct and indirect ownership relations. For the integrated ownership of the external shareholder, given by the row vector $\tilde{d}$, an equation similar to Eq. (2.3) holds

$$\tilde{d} = d + \tilde{d}W^G, \tag{2.7}$$

with the solution

$$\tilde{d} = d(I - W^G)^{-1}. \tag{2.8}$$

Note that $\tilde{d} = (I - [W^G]^t)^{-1}d^t$, where the $t$ denotes the transposition operation. From Eqs. (2.4) and (2.8) the following duality relation can be derived, relating the group value of the external shareholder to its integrated ownership

$$dv^G = \tilde{d}v. \tag{2.9}$$

In Brioschi et al. (1989) this is interpreted as follows: the value of the external shareholder's direct group value portfolio $dv^G$ is equivalent to the integrated portfolio of the underlying values $\tilde{d}v$. This means that the entanglement of ownership relations present in a business group, as seen by an external shareholder, can either be accounted for by considering all direct and indirect links and the firms original values $v_j$ or by taking the direct portfolio using the group values $v_j^G$.

### 2.2.2 *Application to Ownership Networks*

What happens when the external shareholder is himself part of the business group. I.e., how can this model be generalize in the case of arbitrary ownership networks to apply to all nodes?

Note that Eq. (2.7) can be promoted to a matrix equation if we include the external shareholder in the analysis. Without loss of generality, the firms can be ordered in such a way that the adjacency matrix decomposes into the following blocks

$$W = \left(\begin{array}{c|c} 0 & d \\ \hline \vec{0} & W^G \end{array}\right), \tag{2.10}$$

where $\vec{0}$ denotes the column vector containing zeros.

For the group value, Eq. (2.3) simply becomes

$$v^G = W v^G + v. \tag{2.11}$$

Observe that the dimension of $v^G$ of the above equation is one larger than that of $v^G$ of Eq. (2.3), as the external shareholder is now incorporated in the formalism. Furthermore, $v^G$ of Eq. (2.11) is equivalent to Eq. (2.6),

$$v^G = v^{G,\text{tot}}. \tag{2.12}$$

The solution to Eq. (2.11) is

$$v^G = (I - W)^{-1} v. \tag{2.13}$$

For the integrated ownership, Eq. (2.7) is promoted to an operator equation

$$\widetilde{W} = W + \widetilde{W} W. \tag{2.14}$$

The matrix of integrated ownership $\widetilde{W}$ can be understood as a recursive computation of all the indirect paths plus all the direct ones in the network. The solution is given by

$$\widetilde{W} = (I - W)^{-1} W. \tag{2.15}$$

Observe that because for any matrix $A$

$$(I - A)^{-1} = I + A + A^2 + A^3 + \dots \tag{2.16}$$

the following equation holds

$$(I - W)^{-1} W = W (I - W)^{-1}. \tag{2.17}$$

Hence

$$\widetilde{W} = W + W \widetilde{W} \tag{2.18a}$$
$$= W + \widetilde{W} W. \tag{2.18b}$$

Or, as an example of Eq. (2.18a), in scalar form

$$\widetilde{W}_{ij} = W_{ij} + \sum_k W_{ik} \widetilde{W}_{kj}. \tag{2.19}$$

This symmetry seen in Eq. (2.18) in the definition of integrated ownership allows for two equivalent computations. Figure 2.2 gives an illustration of the first recursive definition. To exemplify, take a loop of four firms, where $1 \to 2$, $2 \to 3$, $3 \to 4$,

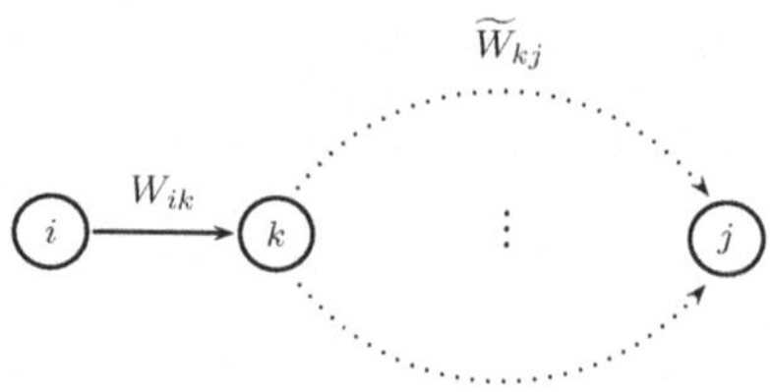


**Fig. 2.2** A schematic illustration of integrated ownership as defined in Eq. (2.19): $\widetilde{W}_{ij}$ is composed of all direct links from firm $i$ to $j$ plus all the direct links (if existent) from $i$ to its neighbors $k$, times the indirect paths of ownership from $k$ to $j$

all links having the weight 0.5, and $4 \to 1$ with the weight 0.1. For the integrated ownership relation $\widetilde{W}_{11}$, Eq. (2.19) can be understood as the direct link $W_{11} = 0$ plus $\sum_k W_{1k}\widetilde{W}_{k1} = W_{12}\widetilde{W}_{21}$. In other words, the direct weighted link from $1 \to 2$ times all the indirect paths from $2 \to 1$. Numerically, $W_{12}\widetilde{W}_{21} = 0.5 \cdot 0.0253 = 0.0126$. Equation (2.18b) reinterprets $\widetilde{W}_{11}$ as the indirect paths from $1 \to 4$ times the direct path from $4 \to 1$: $\widetilde{W}_{14}W_{41} = 0.1266 \cdot 0.1 = 0.0126$.

For the matrix $(I - W)$ to be non-negative and non-singular, a sufficient condition is that the Perron-Frobenius root is smaller than one, $\lambda_{PF}(W) < 1$. A way to see this is by employing the Perron-Frobenius theorem, described in Appendix B.4: for any other eigenvalue $\lambda$ of $W$, $|\lambda| < \lambda_{PF}$. Moreover, for the eigenvector $v_{PF}$, with $Wv_{PF} = \lambda_{PF}v_{PF}$, it holds that

$$(I - W)^{-1}v_{PF} = Iv_{PF} + Wv_{PF} + W^2v_{PF} + \cdots = \frac{1}{1 - \lambda_{PF}}v_{PF}, \tag{2.20}$$

which approaches infinity for $\lambda_{PF} \to 1$. Note that the last relation employed Eq. (2.16).

This condition is ensured by the following requirement: in each strongly connected component $\mathcal{S}$ there exists at least one node $j$ such that $\sum_{i \in \mathcal{S}} W_{ij} < 1$. Hereafter the term "strongly connected component" will be referred to as SCC.[2] In an economics setting, this means that there exists no subset of $k$ firms ($k = 1, \ldots, n$) that are entirely owned by the $k$ firms themselves. A condition which is claimed to be always fulfilled in ownership networks (Brioschi et al. 1989).

The duality relation of Eq. (2.9) now reads

$$Wv^G = \widetilde{W}v. \tag{2.21}$$

As a result, the group value can be understood as

[2] A list of acronyms can be found in front matter of this book.

$$v^G = Wv^G + v \tag{2.22a}$$
$$= \widetilde{W}v + v. \tag{2.22b}$$

In other words, the group value of a firm is not only the sum of the direct ownership percentages in the neighboring companies times their group values plus the firm's own value, but, equivalently, the integrated ownership percentages times the underlying values plus the intrinsic value of the firm itself.

### 2.2.3 Example A

To illustrate all the above introduced concepts, consider the network given in Fig. 2.3. The group's adjacency matrix is

$$W^G = \begin{pmatrix} 0.0 & 0.3 & 0.2 \\ 0.4 & 0.0 & 0.8 \\ 0.4 & 0.7 & 0.0 \end{pmatrix}, \tag{2.23}$$

and the vector of direct ownership relations of the external shareholder is

$$d = \begin{pmatrix} 0.2 & 0.0 & 0.0 \end{pmatrix}. \tag{2.24}$$

Hence the adjacency matrix of the whole network is given by

$$W = \left(\begin{array}{c|ccc} 0.0 & 0.2 & 0.0 & 0.0 \\ \hline 0.0 & 0.0 & 0.3 & 0.2 \\ 0.0 & 0.4 & 0.0 & 0.8 \\ 0.0 & 0.4 & 0.7 & 0.0 \end{array}\right). \tag{2.25}$$

The matrix of integrated ownership is computed from Eq. (2.15) as

$$\widetilde{W} = (I - W)^{-1}W = \begin{pmatrix} 0.00 & 1.00 & 1.00 & 1.00 \\ 0.00 & 4.00 & 5.00 & 5.00 \\ 0.00 & 8.18 & 9.45 & 10.00 \\ 0.00 & 7.73 & 9.32 & 9.00 \end{pmatrix}. \tag{2.26}$$

The vector of group value is

$$v^G = \begin{pmatrix} 4.00 \\ 15.00 \\ 28.64 \\ 27.05 \end{pmatrix}, \tag{2.27}$$

if all $v_i = 1$. The duality relation reads

$$dv^G = 3 = \tilde{d}v. \tag{2.28}$$

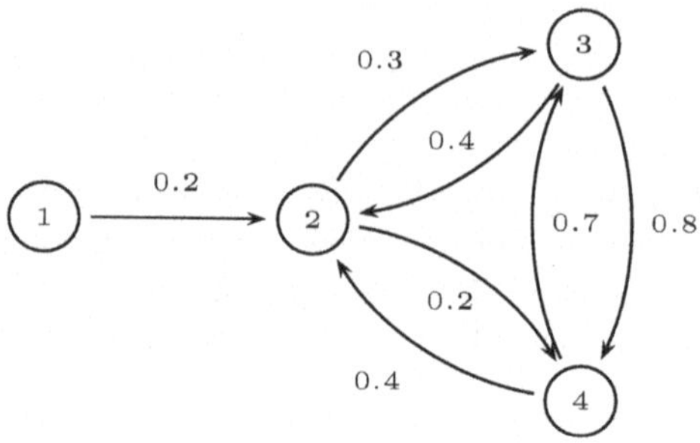


**Fig. 2.3** Simple network example: the external shareholder 1 holds 20 % of the shares of firm 2, which, itself being part of a business group of interconnected companies, has cross-shareholdings in firms 3 and 4

### *2.2.4 Integrated Ownership: Refinements*

Looking at the matrix given in Eq. (2.25), it is natural to consider the generalization where the external shareholder is also part of the business group. I.e., this shareholder also has incoming links. Formally, $W_{1j} \neq 0$ for a certain $j$.

It was, however, soon realized that the presence of self-loops (of any length) is generally problematic. As an example, if firm $i$ owns shares of firm $j$ which in turn owns shares in firm $i$, $i$ owns a portion of itself. But this path of ownership is visited infinitely many times: $i$ also holds shares of itself via the ownership link $i \rightarrow j \rightarrow i \rightarrow j \rightarrow i$, and so forth. This leads to a problem with the economic interpretation of the group value, which grows rapidly when the number of inter-firm cross-shareholdings grows. In effect, the computation overestimates the group value in the presence of SCCs in the network. This is very undesirable behavior and represents a big drawback of the model.

In the example of the last section, the network has many cross-shareholdings between firms 2, 3 and 4, as seen in Fig. 2.3. Looking at their group value given in Eq. (2.27), the smallest is found to be $v_2^G = 15$, although the total value of all three firms connected by cross-shareholdings is only $3 = \sum_i v_i$.

The problem originates from the recursive definition of integrated ownership, as seen in Eq. (2.18). Namely, the components $\widetilde{W}_{ii}$. A solution to this has been proposed in Baldone et al. (1998). The remedy is given by modifying the formalism to remove self-loops of firms connected through cross-shareholdings.

Following the clearer notation of Rohwer and Pötter (2005), Eq. (2.18b) is adapted as follows:

$$\widehat{W}_{ij} = W_{ij} + \sum_{k \neq i} \widehat{W}_{ik} W_{kj}. \tag{2.29}$$

This means that

$$\widehat{W}_{ij} \equiv \widetilde{W}_{ij} - \widetilde{W}_{ii} W_{ij}, \tag{2.30}$$

where $\widetilde{W}_{ii}$ represent all cycles of indirect ownership originating and ending in node $i$. Hence Eq. (2.29) removes incoming links of node $i$ when computing its share of integrated ownership over $j$.

Note that the computations following below are easier to understand when the definition of Eq. (2.18b) is chosen, than when starting with Eq. (2.18a), although both equations are equivalent. In matrix notation, Eq. (2.29) can be manipulated to read

$$\widehat{W} = \left(I - \text{diag}(\widehat{W})\right) W + \widehat{W} W, \tag{2.31}$$

where diag$(A)$ is the matrix of the diagonal of the matrix $A$. In Baldone et al. (1998) the solution is found to be

$$\widehat{W} = \text{diag}(V)^{-1}(V - I), \tag{2.32}$$

defining the quantity

$$V := (I - W)^{-1}. \tag{2.33}$$

This can be re-expressed in scalar form as

$$\widehat{W}_{ij} = \frac{V_{ij}}{V_{ii}}; \quad i \neq j, \tag{2.34a}$$

$$\widehat{W}_{kk} = \frac{V_{kk} - 1}{V_{kk}}. \tag{2.34b}$$

Finally, it can derived that

$$V - I = (I + W + W^2 + \ldots) - I = (I + W + W^2 + \ldots)W = (I - W)^{-1} W = \widetilde{W}, \tag{2.35}$$

using Eq. (2.16).

### 2.2.5 *Modifying the Group Value*

The corrections introduced in the last section will impact the computation of the group value. Baldone et al. (1998) compute the following. Rearranging Eq. (2.31) to read

$$\left(I - \text{diag}(\widehat{W})\right)^{-1} \widehat{W} = W(I - W)^{-1}, \tag{2.36}$$

and multiplying with $v$, yields

$$\left(I - \mathrm{diag}(\widehat{W})\right)^{-1} \widehat{W} v = W(I - W)^{-1} v \tag{2.37a}$$
$$= W v^G \tag{2.37b}$$
$$= v^G - v. \tag{2.37c}$$

Recall the solution of $v^G$ given in Eq. (2.13) and that from Eq. (2.11) the last relation of Eq. (2.37c) can be derived. Finally, Eq. (2.37) is found to be

$$v^G = \left(I - \mathrm{diag}(\widehat{W})\right)^{-1} \widehat{W} v + v. \tag{2.38}$$

By replacing

$$v = \left(I - \mathrm{diag}(\widehat{W})\right) \left(I - \mathrm{diag}(\widehat{W})\right)^{-1} v, \tag{2.39}$$

in Eq. (2.38) and rearranging terms, it can be seen that

$$v^G = \left(I - \mathrm{diag}(\widehat{W})\right)^{-1} \left(\widehat{W} - \mathrm{diag}(\widehat{W})\right) v + \left(I - \mathrm{diag}(\widehat{W})\right)^{-1} v \tag{2.40a}$$
$$= \left(I - \mathrm{diag}(\widehat{W})\right)^{-1} \left((\widehat{W} - \mathrm{diag}(\widehat{W}))v + v\right). \tag{2.40b}$$

Expressed in scalar form, this equation reads

$$v_k^G = \frac{1}{1 - \widehat{W}_{kk}} \left( \sum_{i=1}^{n} \widehat{W}_{ki} v_i - \widehat{W}_{kk} v_k + v_k \right) \tag{2.41a}$$
$$= \frac{1}{1 - \widehat{W}_{kk}} \left( \sum_{i=1, i \neq k}^{n} \widehat{W}_{ki} v_i + v_k \right). \tag{2.41b}$$

The economics interpretation given in Baldone et al. (1998) is as follows. The self-cycles of integrated ownership $\widehat{W}_{kk}$ are understood as referring to treasury shares. These are portions of shares that a company keeps in their own treasury.[3] If there are no treasury shares ($\widehat{W}_{kk} = 0$) and hence no loops back to $k$ in the network, this is equivalent to the external shareholder case covered in Sects. 2.2.1 and 2.2.2. As mentioned, this means that the group value is the sum of the integrated ownership percentages times the values—or the direct ownership shares times the neighbors group values, cf. Eq. (2.22)—plus the underlying value of $k$.

If, however, $\widehat{W}_{kk} > 0$, the group value of firm $k$ exceeds the sum of the (modified) integrated ownership percentages times the values plus $v_k$ by the term $1/(1 - \widehat{W}_{kk})$,

[3] Treasury shares may have come from a repurchase of shares by the firm from shareholders or they may have never been issued to the public. These shares do not pay dividends and have no voting rights, see also Fig. 1.4.

seen in Eq. (2.41). The closer $\widehat{W}_{kk}$ is to one, the greater the group value and the bigger the divergence is.

It is imperative to note the peculiarities of this proposed solution and interpretation. First of all, although the authors in Baldone et al. (1998) correctly identify the problematic term and isolate it in Eq. (2.41), they still do not actually propose a correction to the group value that would result in smaller numerical $v_k^G$ values in the case of cycles in the network. Secondly, in the interpretation of Eq. (2.41b), the modification of integrated ownership the authors propose is very mysterious, namely $\widehat{W} - \text{diag}(\widehat{W})$, as seen in Eq. (2.40b). Especially as a term $\text{diag}(\widehat{W})$ is already present in the definition of $\widehat{W}$, cf. Eq. (2.31).

Our first contributions will be to propose a straightforward correction to the group value in the case of self-loops and a clear interpretation thereof in the next sections. Note that the corrections to the integrated ownership proposed above are still being used, see for instance (Rohwer and Pötter 2005; Chapelle 2005). Indeed, even the uncorrected methodology is still in use (Almeida et al. 2007).

## 2.3 Introducing Network Value and Integrated Value

Building on the studies of Brioschi et al. (1989), Baldone et al. (1998) we present an extension of the formalism and remedy the shortcomings which have been unaddressed to this date (see the end of the last section for details). Moreover, there is a very elegant interplay between the notions of integrated ownership and group value which has never been explicitly pointed out in the literature.

### *2.3.1 Integrated Value*

In analogy to the notion of the portfolio value of shareholders defined in Eq. (1.2), we define the *integrated value*

$$\tilde{\nu}^{\text{int}} := \widetilde{W} v = (I - W)^{-1} W v. \tag{2.42}$$

$\tilde{\nu}_i^{\text{int}}$ represents the portfolio value of shareholder $i$ considering all its direct and indirect (i.e., integrated) paths of ownership in the network. Recall that the integrated ownership matrix obeys the operator equation

$$\widetilde{W}_{ij} = W_{ij} + \sum_k W_{ik} \widetilde{W}_{kj}. \tag{2.43}$$

as seen in Eq. (2.19) and Fig. 2.2. It should be noted, that in the spirit of the 3-level network analysis mentioned in Sect. 1.1.1, $\tilde{\nu}_i^{\text{int}}$ is a fully fledged Level 3 measure, incorporating all the available information of the complex network under study.

Observe that the integrated ownership value of Eq. (2.42) is also the solution to the following equation

$$\tilde{\nu}^{\text{int}} = W\tilde{\nu}^{\text{int}} + Wv, \tag{2.44}$$

or in scalar notation

$$\tilde{\nu}_i^{\text{int}} = \sum_k W_{ik}\tilde{\nu}_k^{\text{int}} + \sum_k W_{ik}v_k, \tag{2.45}$$

which can be interpreted as a centrality measure similarly to Eqs. (2.3) and (2.11), which are reminiscent of a Hubbell index centrality (see Sect. 2.6). In effect, the importance of node $i$ reflected in $\tilde{\nu}_i^{\text{int}}$, is determined by the importance of its neighbors and the value of its neighbors. Alternatively, in an ownership setting, $\tilde{\nu}_i^{\text{int}}$ in Eq. (2.45) should be understood as the integrated value of $i$'s neighbors plus $i$'s portfolio value, given in Eq. (1.2). An additional interpretation in terms of network flow will be given in Sect. 2.4.

Although $\tilde{\nu}^{\text{int}}$ was implicitly used in the duality relation of Eq. (2.21), the context given in Eq. (2.44) was previously unobserved in the pertinent literature.

### 2.3.2 Network Value

In the rest of this thesis, the term *network value* will be used to replace the notion of group value introduced in Sect. 2.2.1. The change in naming reflects the fact that it is a general network measure and not necessarily constrained to the idea of business groups.

To generalize Eq. (2.11)

$$v^{\text{net}} := Wv^{\text{net}} + v, \tag{2.46}$$

and

$$v^{\text{net}} = (I - W)^{-1}v. \tag{2.47}$$

In an ownership context, Eq. (2.46) can be understood as computing $i$'s network value as the direct portfolio of its neighbors network value plus $i$'s own underlying value $v_i$.

### 2.3.3 The Whole Picture

For the duality relation of Eq. (2.21) one can easily see that the following relations hold, employing Eq. (2.17)

$$\widetilde{W}v = (I - W)^{-1}Wv = W(I - W)^{-1}v = Wv^{\text{net}}. \tag{2.48}$$

Finally, combining Eqs. (2.46) and (2.48) uncovers the novel connection between the two concepts

$$v^{\text{net}} = Wv^{\text{net}} + v = \widetilde{W}v + v = \tilde{v}^{\text{int}} + v. \tag{2.49}$$

In other words, the network value accounts for the overall value of an economic actor, given by its underlying value plus the value gained from the integrated value. Moreover, the integrated value reflects the value attained from the underlying values of all firms reachable by all direct and indirect paths of ownership. It is also equivalent to the network value of all directly owned firms.

### 2.3.4 The True Corrections

As mentioned at the end of Sect. 2.2.5, the corrections proposed in Baldone et al. (1998) have not been implemented correctly. In order to understand this, it is crucial to reformulate Eq. (2.31) appropriately. This is best done by introducing the *correction operator*

$$\mathcal{D} := \text{diag}\left((I - W)^{-1}\right)^{-1}, \tag{2.50}$$

or using Eq. (2.33)

$$\mathcal{D} = \text{diag}(V)^{-1}. \tag{2.51}$$

Recall that diag$(A)$ is defined as the matrix of the diagonal elements of the matrix $A$. The components of $\mathcal{D}$ are

$$\mathcal{D}_{kk} = \frac{1}{(I - W)^{-1}_{kk}}, \tag{2.52a}$$

$$\mathcal{D}_{ij} = 0, \quad i \neq j. \tag{2.52b}$$

To express $\widehat{W}$ in terms of $\mathcal{D}$ one can insert Eq. (2.51) directly into Eq. (2.32), recalling Eq. (2.35). Or, alternatively, one can start from the defining relation given in Eq. (2.31). One then must first derive the following algebraic identity using the properties of the diag() operation[4]

[4] $\text{diag}(\text{diag}(A)) = \text{diag}(A)$, $\text{diag}(A + B) = \text{diag}(A) + \text{diag}(B)$, and $\text{diag}(I) = I$.

$$\begin{aligned}\mathrm{diag}(\widehat{W}) &= \mathrm{diag}\left(\mathrm{diag}(V)^{-1}(V-I)\right) = \mathrm{diag}(V)^{-1}\mathrm{diag}\,(V-I) \\ &= \mathrm{diag}(V)^{-1}\,(\mathrm{diag}(V)-I) = I - \mathrm{diag}(V)^{-1} = I - \mathcal{D},\end{aligned} \tag{2.53}$$

in other words

$$\mathcal{D} = I - \mathrm{diag}(\widehat{W}). \tag{2.54}$$

Inserting this directly into Eq. (2.31) and solving for $\widehat{W}$ reveals the wanted relation. To summarize, both possibilities yield

$$\widehat{W} = \mathcal{D}W(I-W)^{-1} = \mathcal{D}(I-W)^{-1}W = \mathcal{D}\widetilde{W}. \tag{2.55}$$

It has now become apparent that the effect of the correction to the integrated ownership due to self-loops proposed by Baldone et al. (1998) in Eq. (2.29) is equivalent to the simple multiplication of $\mathcal{D}$ and $\widetilde{W}$.

This now allows us to define the corrected integrated value as

$$\hat{\nu}^{\mathrm{int}} := \widehat{W}v = \mathcal{D}\widetilde{W}v = \mathcal{D}\tilde{\nu}^{\mathrm{int}}, \tag{2.56}$$

recalling Eq. (2.42).

In a similar vein we introduce the missing correction to the network value. Starting from where Baldone et al. (1998) left off, namely Eq. (2.41b), we propose the following interpretation

$$v_k^{\mathrm{net}} = \frac{1}{1-\widehat{W}_{kk}}\left(\sum_{i=1, i\neq k}^{n} \widehat{W}_{ki}v_i + v_k\right) \tag{2.57a}$$

$$= \frac{1}{1-\widehat{W}_{kk}}\left(\sum_{i=1}^{n} \widehat{W}_{ki}v_i + (1-\widehat{W}_{kk})v_k\right). \tag{2.57b}$$

Now Eq. (2.57b) can be re-expressed in matrix notation as

$$\begin{aligned} v^{\mathrm{net}} &= \left(I - \mathrm{diag}(\widehat{W})\right)^{-1}\left(\widehat{W}v + \left(I-\mathrm{diag}(\widehat{W})\right)v\right) \\ &= \mathcal{D}^{-1}\left(\mathcal{D}\widetilde{W}v + \mathcal{D}v\right),\end{aligned} \tag{2.58}$$

employing Eq. (2.53). Hence a natural interpretation of the impact of the loop-correction on the network value is

$$\hat{v}^{\mathrm{net}} := \mathcal{D}v^{\mathrm{net}} = \mathcal{D}(\tilde{\nu}^{\mathrm{int}} + v) = \mathcal{D}\widetilde{W}v + \mathcal{D}v = \widehat{W}v + \mathcal{D}v = \hat{\nu}^{\mathrm{int}} + \mathcal{D}v. \tag{2.59}$$

The introduction of the modified network value $\hat{v}^{\mathrm{net}} = \mathcal{D}v^{\mathrm{net}}$ is in complete analogy to the corrected integrated value $\hat{\nu}^{\mathrm{int}} = \mathcal{D}\tilde{\nu}^{\mathrm{int}}$, seen in Eq. (2.56). To summarize,

**Table 2.1** Summary of integrated value $\tilde{\nu}^{\text{int}}$, network value $\upsilon^{\text{net}}$ and their correction $\hat{\nu}^{\text{int}}$ and $\hat{\upsilon}^{\text{net}}$

| Definition | Solution | Self-loop-correction |
|---|---|---|
| $\tilde{\nu}^{\text{int}} = W(\nu^{\text{int}} + \upsilon)$ | $\tilde{\nu}^{\text{int}} = \widetilde{W}\upsilon$ | $\hat{\nu}^{\text{int}} = \mathcal{D}\widetilde{W}\upsilon$ |
| $\upsilon^{\text{net}} = W\upsilon^{\text{net}} + \upsilon$ | $\upsilon^{\text{net}} = \tilde{\nu}^{\text{int}} + \upsilon$ | $\hat{\upsilon}^{\text{net}} = \hat{\nu}^{\text{int}} + \mathcal{D}\upsilon$ |

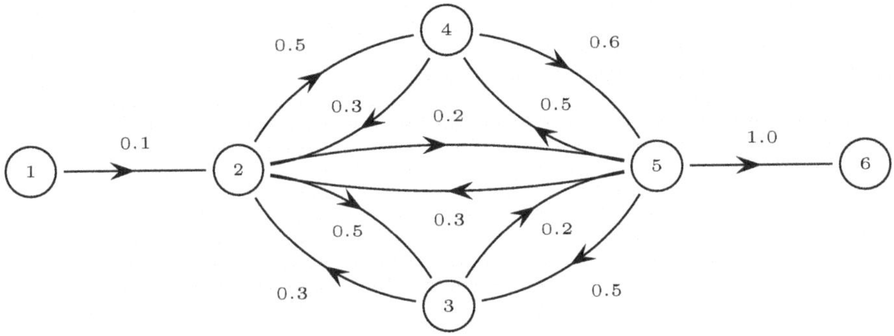


**Fig. 2.4** Simple bow-tie network topology example with a high degree of interconnectedness of the firms in the strongly connected component (SCC)

the effect of removing the incoming links of a firm $i$ in the analysis results in the underlying value $\upsilon_i$, the network value $\upsilon_i^{\text{net}}$, and the integrated value $\tilde{\nu}^{\text{int}}$ all being multiplied by a factor $\mathcal{D}_{ii} = 1 - \widehat{W}_{ii} = 1/(I - A)^{-1}_{ii}$:

$$\mathcal{D}\upsilon^{\text{net}} = \mathcal{D}\tilde{\nu}^{\text{int}} + \mathcal{D}\upsilon. \tag{2.60}$$

This also underlines the crucial role played by the correction operator $\mathcal{D}$, which, by incorporating all the effects of amending for self-loops, acts as a consistent measure to derive the corrected terms.

Table 2.1 summarizes all the important relations that have been derived so far. In the next section, a numerical example is presented.

### 2.3.5 *Example B: And the Next Problem on the Horizon*

Consider the network illustrated in Fig. 2.4. It is an example of a simple bow-tie network topology. The SCC is constructed in a way to highlight the problem of cross-shareholdings. Hence there are many cycles of indirect ownership originating and ending in each firm in the core of the bow-tie.

We assume the underlying value of each firm to be one, i.e., $\upsilon = (1, 1, 1, 1, 1, 1)^t$, where $t$ denotes the transposition operation. This results in the network value and the integrated value to be

$$v^{\text{net}} = \begin{pmatrix} 6 \\ 50 \\ 27 \\ 49 \\ 55 \\ 1 \end{pmatrix}, \qquad \tilde{v}^{\text{int}} = \begin{pmatrix} 5 \\ 49 \\ 26 \\ 48 \\ 54 \\ 0 \end{pmatrix}. \tag{2.61}$$

So although the total value present in the network is $6 = \sum_i v_i$, firm 5 has an disproportionately large network value of $v_5^{\text{net}} = 55$.

Introducing the correction operator, one finds

$$\mathcal{D} = \begin{pmatrix} 1.000 & 0 & 0 & 0 & 0 & 0 \\ 0 & 0.100 & 0 & 0 & 0 & 0 \\ 0 & 0 & 0.162 & 0 & 0 & 0 \\ 0 & 0 & 0 & 0.095 & 0 & 0 \\ 0 & 0 & 0 & 0 & 0.086 & 0 \\ 0 & 0 & 0 & 0 & 0 & 1.000 \end{pmatrix}, \tag{2.62}$$

allowing the corrected values to be computed as

$$\hat{v}^{\text{net}} = \begin{pmatrix} 6.000 \\ 5.000 \\ 4.378 \\ 4.667 \\ 4.714 \\ 1.000 \end{pmatrix}, \qquad \hat{v}^{\text{int}} = \begin{pmatrix} 5.000 \\ 4.900 \\ 4.216 \\ 4.571 \\ 4.629 \\ 0.000 \end{pmatrix}. \tag{2.63}$$

The correction reduces the values of the firms in the core of the bow-tie by approximately one order of magnitude. This confirms that $\hat{v}^{\text{net}}$ and $\tilde{v}^{\text{int}}$ are indeed the right measures to consider in the presence of SCCs in the network.

Unfortunately, this example highlights a further problem of the methodology. We will present two solutions to remedy this complication in Sect. 2.5. Before detailing the issue, it should be noted that the problem at hand was previously overlooked, because the ownership networks that were analyzed did not have a bow-tie structure and because the focus was not on the empirical analysis of control. It plagues both $\tilde{v}^{\text{int}}$ and $\hat{v}^{\text{int}}$.

In a nutshell, the problem can be described as follows: a single root node $r$, i.e., $k_r^{in} = 0$, connected to a SCC will be assigned an integrated value which is the sum of the underlying value of all the firms reachable by ownership paths from $r$. In the above example, $\tilde{v}_1^{\text{int}} = \hat{v}_1^{\text{int}} = 5$. This behavior is, however, independent of the percentage of ownership connecting $r$ to the core, e.g., $W_{12}$ in Fig. 2.4. This means that a company with no shareholders and an arbitrarily small share in a firm in the SCC (having no other external shareholders) still gets an integrated value totalling the underlying value of all firms it has integrated ownership in. This is obviously a very undesirable behavior. Note that if the SCC has multiple root-nodes connecting to it, the total underlying value gets distributed amongst them, also regardless of the link strength.

In order to fully understand this intricacy, a short digression into the theory of networks is necessary. In detail, the idea of a quantity flowing in the network gives an alternative interpretation of integrated value and network value. This change in point of view facilitates the understanding of the above mentioned problem.

## 2.4 A New Perspective: The Notion of Flow in Networks

Consider a directed and weighted network in which (i) a non-topological real value $v_j \geq 0$ can be assigned to the nodes (with the condition that $v_j > 0$ for at least all the leaf-nodes in the network, i.e., nodes with $k_i^{out} = 0$) and (ii) an edge from node $i$ to $j$ with weight $W_{ij}$ implies that some of the value of $j$ is transferred to $i$. In terms of a physical system, we think of the nodes as entities receiving material from the downstream nodes and transferring it to the upstream nodes, without dissipation, in proportion to the weights of the incoming links.

Assume that the nodes which are associated with a value $v_j$ produce $v_j$ units of mass (or energy) at time $t = 1$. Then the flow $\phi_i$ entering the node $i$ from each node $j$ at time $t$ is the fraction $W_{ij}$ of the mass produced directly by $j$ plus the same fraction of the inflow of $j$:

$$\phi_i(t+1) = \sum_j W_{ij} v_j + \sum_j W_{ij} \phi_i(t). \tag{2.64}$$

where $\sum_i W_{ij} = 1$ for the nodes that have predecessors and $\sum_i W_{ij} = 0$ for the root-nodes (sinks). In matrix notation, at the steady state ($t \to \infty$), this yields

$$\phi = W(v + \phi). \tag{2.65}$$

The solution

$$\phi = (1 - W)^{-1} W v, \tag{2.66}$$

exists and is unique if $\lambda(W) < 1$. This condition is easily fulfilled in real networks as it requires that in each SCC $\mathcal{S}$ there exists at least one node $j$ such that $\sum_{i \in \mathcal{S}} W_{ij} < 1$. Or, equivalently, the mass circulating in $\mathcal{S}$ is also flowing to some node outside of $\mathcal{S}$. Notice that this does not imply that mass is lost in the transfer. Indeed, the mass is conserved at all nodes except at the sinks. Some of the nodes only produce mass (all the leaf-nodes but possibly also other nodes) at time $t = 1$ and are thus sources, while the root-nodes accumulate the mass.

Note that it is straightforward to also define an equation for the evolution of the stock of mass (energy) present at each node. The convention used here implies that mass flows against the direction of the edges. The transported quantity is only created once at time $t = 1$ and the root nodes get self-loops assigned to them, so that no quantity is lost. For non-root nodes (i.e., $k_i^{in} > 0$) the stock never gets accumulated and is always passed on upstream. For the root nodes, or sinks, the value for the flow and the accumulated stock is equivalent.

The equation for the evolution of the stock $s$ of mass present at each node can be derived as follows: for node $i$ the stock at time $t + 1$ is the value of the previous time step minus outflow plus inflow

$$s_i(t+1) := s_i(t) - \sum_k W_{ki} s_i(t) + \sum_j W_{ij} s_j(t), \tag{2.67}$$

or equivalently

$$s_i(t+1) = \begin{cases} s_i(t) + \sum_j W_{ij} s_j(t), & \text{if } i \text{ is a sink,} \\ \sum_j W_{ij} s_j(t), & \text{otherwise.} \end{cases} \tag{2.68}$$

In matrix notation the equation above reads

$$s(t+1) := \mathcal{T} s(t) = (\mathcal{S} + W) s(t), \tag{2.69}$$

where

$$\mathcal{S}_{ii} = 1, \quad \text{if } i \text{ is a sink,} \tag{2.70a}$$

$$\mathcal{S}_{ij} = 0, \quad \text{otherwise.} \tag{2.70b}$$

In effect, the diagonal matrix $\mathcal{S}$ assigns self-loops to sinks, conserving the mass or energy in the network.

As a result of Eq. (2.69)

$$s(n) = \mathcal{T}^n s(0). \tag{2.71}$$

Note that $\mathcal{T}^n$ describes paths of length $n$ in the network. However, does Eq. (2.71) converge, meaning that after some time $\hat{t}$ the stock is unchanged: $s(\hat{t}+1) = s(\hat{t})$? To see that there exists a fixed point $s_*$

$$s_* = \mathcal{T} s_*, \tag{2.72}$$

consider the following. The ownership matrix $W$ is per definition non-negative and column stochastic, i.e., $\sum_j W_{ij} \leq 1$ and, per construction, $W_{ii} = 0$. Hence $\mathcal{T}$ is also non-negative and column stochastic. This means that the Perron-Frobenius theorem, explained in Appendix B.4, holds and there exists a unit eigenvalue of $\mathcal{T}$: $\lambda = 1$. In other words

$$\mathcal{T} s_* = \lambda s_* = s_*. \tag{2.73}$$

## 2.4.1 Flow in Ownership Networks

So what does this all mean in the case of an ownership network? And what quantity can be seen as flowing along the links? The cash allowing an equity stake in a firm to be held flows in the direction of the edges. In contrast, the ownership of a firm's equity capital, i.e., the cash-flow rights, are transferred in the opposite direction, from the firm to its shareholders. The same is true for the payed dividends (and voting rights, see Sect. 2.7). See also Sect. 1.2.1.

Observe that Eq. (2.66) is equivalent to Eq. (2.42), uncovering the following interpretation

$$\phi \equiv \tilde{\nu}^{\text{int}}. \tag{2.74}$$

In other words, the integrated value $\tilde{\nu}_i^{\text{int}}$ in an ownership network corresponds to the inflow $\phi_i$ of the underlying units of value $v_i$ in the steady state.

It is now possible to conceptually understand the problem mentioned at the end of Sect. 2.3.5. Since the integrated value of any node corresponds to the inflow over an infinite time, all the value ($v^{tot} = \sum_i v_i$) that is flowing in the network will ultimately accumulate in the root nodes. In the case of a single root node $r$ connected to a SCC, as in the example given in Fig. 2.4, the total value of all the firms downstream of it will necessarily have to flow to it, regardless of the percentage of ownership with which the root node is connected to firms in the core: $\tilde{\nu}_r^{\text{int}} = \sum_{i \neq r} v_i$.

Finally, for the problem of the overestimation of integrated value, as mentioned in Sect. 2.2.4, the following should be noted. The more indirect self-cycles are present in the network, the longer the quantity will be circulating through the nodes connected by these paths. This explains the high numerical values of $\tilde{\nu}_i^{\text{int}}$ for nodes in SCCs. However, because $\tilde{\nu}^{\text{int}} = \phi$, these are actually the correct value to assign to such nodes in a physical network. Only in the context of ownership, this behavior is seen as pathological and $\hat{\nu}^{\text{int}}$ is introduced to alleviate this characteristic. It is therefore important to keep in mind, that although $\hat{\nu}^{\text{int}}$ has a more desirable behavior in the context of ownership networks, it has no correspondence to a physical system anymore.

## 2.5 The Final Corrections: Adjusting Network Value and Integrated Value for Bow-Tie Topologies

To summarize, all previous versions of network value and integrated value failed for ownership networks with bow-tie topologies:

1. firms in the SCC get assigned excessively high quantities;
2. firms with no shareholders accumulate the underlying value of the firms they have integrated ownership in.

As indicated at the end of Sect. 2.3.5, we will now present two related solutions to these problems. The first version will be an analytical derivation of the new quantities. The second solution is given by an algorithm.

For smaller networks, the analytical solution is easily implemented. However, if large networks need to be analyzed, none of the analytical measures are feasible, as already the computation of the inverse matrix $(I - W)$ becomes intractable. Hence the application of the algorithm is inevitable.

### 2.5.1 The Analytical Solution

Observe that the duality relation $\widetilde{W} v = W v^{\text{net}}$, given in Eq. (2.48), is lost for the changes introduced by the correction operator $\mathcal{D}$, defined in Eqs. (2.50) or (2.52b). This is easily seen by

$$\hat{v}^{\text{int}} = \widehat{W} v = \mathcal{D}\widetilde{W} v = \mathcal{D} W v^{\text{net}} \tag{2.75a}$$

$$\neq W\mathcal{D} v^{\text{net}} = W \hat{v}^{\text{net}}. \tag{2.75b}$$

Recall Table 2.1 for a summary of the definitions and relations. In effect, the non-commutative nature of the matrix multiplication, $\mathcal{D}W \neq W\mathcal{D}$, is responsible for the inequality $\widehat{W} v \neq W \hat{v}^{\text{net}}$. The result of this subtlety is that one can define two variants related to this correction. Next to $\hat{v}^{\text{net}} = \widehat{W} v + \mathcal{D} v = \mathcal{D} W v^{\text{net}} + \mathcal{D} v$, cf. Eq. (2.59), the natural alternative is

$$\bar{v}^{\text{net}} := W\mathcal{D} v^{\text{net}} + \mathcal{D} v = W \hat{v}^{\text{net}} + \mathcal{D} v. \tag{2.76}$$

The following algebraic manipulations uncover the meaning of this equation. Replacing $v^{\text{net}}$ with the relation given in Eq. (2.49)

$$\bar{v}^{\text{net}} = W\mathcal{D}\left(\widetilde{W} v + v\right) + \mathcal{D} v = W(\widehat{W} + \mathcal{D}) v + \mathcal{D} v =: \overline{W} v + \mathcal{D} v. \tag{2.77}$$

This identifies an additional corrected integrated ownership matrix as

**Table 2.2** Summary of $\tilde{\nu}^{\text{int}}$, $\upsilon^{\text{net}}$, $\hat{\nu}^{\text{int}}$, $\hat{\upsilon}^{\text{net}}$, $\bar{\nu}^{\text{int}}$, $\bar{\upsilon}^{\text{net}}$, $\mathcal{D}$, $\widetilde{W}$, $\widehat{W}$, and $\overline{W}$

| Definition | Solution | First correction | Second correction |
|---|---|---|---|
| $\tilde{\nu}^{\text{int}} = W(\nu^{\text{int}} + \upsilon)$ | $\tilde{\nu}^{\text{int}} = \widetilde{W}\upsilon$ | $\hat{\nu}^{\text{int}} = \widehat{W}\upsilon$ | $\bar{\nu}^{\text{int}} = \overline{W}\upsilon$ |
| $\upsilon^{\text{net}} = W\upsilon^{\text{net}} + \upsilon$ | $\upsilon^{\text{net}} = \tilde{\nu}^{\text{int}} + \upsilon$ | $\hat{\upsilon}^{\text{net}} = \hat{\nu}^{\text{int}} + \mathcal{D}\upsilon$ | $\bar{\upsilon}^{\text{net}} = \bar{\nu}^{\text{int}} + \mathcal{D}\upsilon$ |
| $\mathcal{D} = \text{diag}\left((I-W)^{-1}\right)^{-1}$, | $\widetilde{W} = (I-W)^{-1}W$, | $\widehat{W} = \mathcal{D}\widetilde{W}$, | $\overline{W} = W(\widehat{W} + \mathcal{D})$. |

$$\overline{W} := WW^*, \tag{2.78}$$

with

$$W^* := \widehat{W} + \mathcal{D}, \tag{2.79}$$

or in scalar notation

$$W^*_{ij} = \begin{cases} 1, & i = j, \\ \widehat{W}_{ij}, & i \neq j. \end{cases} \tag{2.80}$$

In analogy to the previous sections, it is now straightforward to introduce the corresponding corrected integrated value

$$\bar{\nu}^{\text{int}} := \overline{W}\upsilon, \tag{2.81}$$

which identifies $\bar{\upsilon}^{\text{net}}$ as an additional corrected network value

$$\bar{\upsilon}^{\text{net}} = \bar{\nu}^{\text{int}} + \mathcal{D}\upsilon. \tag{2.82}$$

In Table 2.2 all the introduced concepts are summarized. Before we analyze the behavior of $\bar{\upsilon}^{\text{net}}$ and $\bar{\nu}^{\text{int}}$, we first introduce the corresponding algorithmic solution in the next section.

Finally, the following equations give all the equalities related to the various incarnations of network value and integrated value:

$$\upsilon^{\text{net}} = \nu^{\text{int}} + \upsilon = \widetilde{W}\upsilon + \upsilon, \tag{2.83}$$

$$\hat{\upsilon}^{\text{net}} = \hat{\nu}^{\text{int}} + \mathcal{D}\upsilon = \widehat{W}\upsilon + \mathcal{D}\upsilon = \mathcal{D}\widetilde{W}\upsilon + \mathcal{D}\upsilon = \mathcal{D}W\upsilon^{\text{net}} + \mathcal{D}\upsilon, \tag{2.84}$$

$$\bar{\upsilon}^{\text{net}} = \bar{\nu}^{\text{int}} + \mathcal{D}\upsilon = \overline{W}\upsilon + \mathcal{D}\upsilon = W\mathcal{D}\widetilde{W}\upsilon + W\mathcal{D}\upsilon + \mathcal{D}\upsilon = W\mathcal{D}\upsilon^{\text{net}} + \mathcal{D}\upsilon. \tag{2.85}$$

## *2.5.2 The Algorithmic Solution*

We illustrate the algorithm for the computation of the network value. Then the integrated value can be obtained by deducting the underlying value. In order to calculate the network value for any specific node $i$, we extract the whole subnetwork that is downstream of a node $i$, including $i$. For this purpose, a breadth-first-search (BFS) returns the set of all nodes reachable from $i$, going in the direction of the links. Then, all the links among these nodes are obtained from the adjacency matrix of the entire network, except for the links pointing to $i$ which are removed. This ensures that there are no cycles involving $i$ present in the subnetwork. Let $B(i)$ denote the adjacency matrix of such a subnetwork, including $i$, extracted from the ownership matrix $W$. Without loss of generality, we can relabel the nodes so that $i = 1$. Since node 1 has now no incoming links, we can decompose $B = B(1)$ as follows:

$$B = \left(\begin{array}{c|c} 0 & d \\ \hline \vec{0} & B^{\text{sub}} \end{array}\right), \tag{2.86}$$

where $d$ is the vector of all links originating from node 1, and $B^{\text{sub}}$ is associated with the subgraph of the nodes downstream of $i$. This is similar to the decomposition given in Eq. (2.10) for the case with an external shareholder.

The underlying value of these nodes is given by the vector $v^{\text{sub}}$. By replacing the matrix $B$ in the expression $v^{\text{net}} = \widetilde{W}v + v = W(I - W)^{-1}v + v$, cf. Eq. (2.49), and taking the first component we obtain:

$$v^{\text{net}}(1) = \left[B(I - B)^{-1}v\right]_1 + v_1 \tag{2.87a}$$

$$= \left[d(I^{\text{sub}} - B^{\text{sub}})^{-1}v^{\text{sub}}\right]_1 + v_1 =: \tilde{d} \cdot v^{\text{sub}} + v_1, \tag{2.87b}$$

where now $\tilde{v}^{\text{int}}(1) := \tilde{d} \cdot v^{\text{sub}}$, in analogy to the term in Eq. (2.8).

Notice that if node $i$ has zero in-degree, this procedure yields the same result as the previous formula for the integrated ownership matrix of Eq. (2.15): $\tilde{B}_{(i,*)} = (0, \tilde{d}) = \widetilde{W}_{(i,*)}$. The notation $A_{(i,*)}$ for a matrix is understood as taking its $i$-th row. In Appendix D.1 it is formally shown that our calculation is in fact equivalent to the correction proposed by Baldone et al. (1998) to address the problems of the overestimation of network value in the case of ownership due to the presence of cycles.

However, the methods still suffer from the problem of root nodes accumulating all the value flowing in the network. To solve this issue, we adjust our algorithm to pay special attention to the IN-nodes of an SCC. We partition the bow-tie associated with this SCC into its components: the IN (to which we also add the T&T), the SCC itself, and the OUT. Then, we proceed in multiple steps to compute the network value for all parts in sequence. In this way, the value flows from the OUT, via the SCC to the IN. Finally, the integrated value of firm $i$ is computed from the network value as

$$\tilde{v}^{\text{int}}(i) = v^{\text{net}}(i) - v_i. \tag{2.88}$$

In detail, our algorithm works as follows:

1. OUT: Compute the network value $v^{\text{net}}(i)$ for all the nodes in the OUT using Eq. (2.87b).
2. OUT $\rightarrow$ SCC: Identify the subset $\mathcal{S}1$ of nodes in the SCC pointing to nodes in the OUT, the latter subset denoted as $\mathcal{O}$. To account for the value entering the SCC from the OUT, compute the network value of these selected nodes by applying $v^{\text{net}}(s) = \sum_o W_{so} v^{\text{net}}(o) + v_s$ to them. This is an adaptation of Eq. (2.49), where $s$ and $o$ are labels of nodes in $\mathcal{S}1$ and $\mathcal{O}$, respectively. Note that we only needed to consider the direct links for this. This computation is also equivalent to applying Eq. (2.87b), which considers the downstream subnetworks of $\mathcal{S}1$, i.e., the whole OUT.
3. SCC: Employ Eq. (2.87b) to the SCC-nodes restricting the BFS to retrieve only nodes in the SCC itself. Note that for those SCC-nodes that were already considered in step 2, their network value is now taken as the intrinsic value in the computation. This means one first needs to assign $v_i \mapsto v^{\text{net}}(i) + v_i$.
4. SCC $\rightarrow$ IN: In this step we solve the problem of the root-nodes acquiring an exaggerated fraction of the network value. For the subset of IN-nodes $\mathcal{I}$ directly connected to some SCC-nodes $\mathcal{S}2$, we again apply $v^{\text{net}}(i) = \sum_s \mathcal{W}_{is} v^{\text{net}}(s) + v_i$, where $i$ and $s$ are labels of nodes in $\mathcal{I}$ and $\mathcal{S}2$, respectively. However, note that due to the cycles present in the SCC, this computation is not equivalent to Eq. (2.87b). In other words, the duality relation similar to Eq. (2.48), ensuring that the direct portfolio of group value is equivalent to the portfolio of the integrated underlying values, is violated due to the presence of self-loops[5]: $\sum_s \mathcal{W}_{is} v^{\text{net}}(s) \neq \nu^{\text{int}}(i)$. In this way only the direct share of network value over the SCC which is not owned by other SCC-nodes is transfered to the IN-nodes.
5. IN: Finally, use Eq. (2.87b) for assigning the network value to the nodes in the IN-subnetwork. In this case the BFS should not consider the SCC-nodes since their value has been already transfered to their first neighbors in the IN. However, it should retrieve the T&T departing from the IN. Again, for the IN-nodes treated in step 4, first assign $v_i \mapsto v^{\text{net}}(i) + v_i$.

Notice that if any part of the bow-tie structure contains additional smaller SCCs, these should be treated first, by applying steps two to four.

This dissection of the network into its bow-tie components also reduces the computational problems. Although we perform a BFS for each node and compute the inverse of the resulting adjacency matrix of the subnetwork as seen in Eq. (2.87b), the smaller sizes of the subnetworks allow faster computations.

To summarize, the algorithm computes the network value of firm $i$ as $v^{\text{net}}(i)$. By deducting the underlying value, we retrieve the integrated value of $i$: $\tilde{\nu}^{\text{int}}(i)$.

The algorithm presented here is applied to the global network of transnational corporations in Chap. 4.

[5] This was also observed in Eq. (2.75).

### 2.5.3 Revisiting Example B: A Summary and Discussion

Coming back to the network example shown in Fig. 2.4, we now compute all the relevant expressions derived in this chapter and discuss the results. In the following, to avoid any confusion, the row vector of network values $\upsilon^{\text{net}}(i)$, computed from the algorithmic solution as detailed in Eq. (2.87), will be identified as

$$\overset{\circ}{\upsilon}{}_i^{\text{net}} := \upsilon^{\text{net}}(i). \tag{2.89}$$

Recall that the algorithmic computation of network value must be performed for each node separately. Hence the vector $\overset{\circ}{\upsilon}{}^{\text{net}}$ requires Eq. (2.87) to have been computed $n$ times, if $n$ is the length of $\overset{\circ}{\upsilon}{}^{\text{net}}$. Although this appears to be rather tedious, it actually makes the algorithm applicable for very large networks, as observed at the end of the last section.

Correspondingly, for the integrated value,

$$\overset{\circ}{\nu}{}_i^{\text{int}} := \tilde{\nu}^{\text{int}}(i) = \overset{\circ}{\upsilon}{}_i^{\text{net}} - \upsilon_i, \tag{2.90}$$

will denote the integrated value computed from the algorithm. Note that this relation stems from Eq. (2.87).

Recall that the underlying values are chosen to be

$$\upsilon = (1, 1, 1, 1, 1, 1)^t . \tag{2.91}$$

It follows that

$$\mathcal{D}\upsilon = (1.000, 0.100, 0.162, 0.095, 0.086, 1.000)^t , \tag{2.92}$$

using $\mathcal{D}$ from Eq. (2.62). As usual, $t$ denotes the transposition operation.

The corresponding measures of network value are

$$\upsilon^{\text{net}} = \begin{pmatrix} 6 \\ 50 \\ 27 \\ 49 \\ 55 \\ 1 \end{pmatrix}, \quad \hat{\upsilon}^{\text{net}} = \begin{pmatrix} 6.000 \\ 5.000 \\ 4.378 \\ 4.667 \\ 4.714 \\ 1.000 \end{pmatrix}, \quad \overset{\circ}{\upsilon}{}^{\text{net}} = \begin{pmatrix} 1.500 \\ 5.000 \\ 4.378 \\ 4.667 \\ 4.714 \\ 1.000 \end{pmatrix}, \quad \bar{\upsilon}^{\text{net}} = \begin{pmatrix} 1.500 \\ 5.565 \\ 2.605 \\ 4.424 \\ 7.108 \\ 1.000 \end{pmatrix}. \tag{2.93}$$

The quantities of integrated value are

$$\tilde{\upsilon}^{\text{int}} = \begin{pmatrix} 5 \\ 49 \\ 26 \\ 48 \\ 54 \\ 0 \end{pmatrix}, \; \hat{\upsilon}^{\text{int}} = \begin{pmatrix} 5.000 \\ 4.900 \\ 4.216 \\ 4.571 \\ 4.629 \\ 0.000 \end{pmatrix}, \; \mathring{\upsilon}^{\text{int}} = \begin{pmatrix} 0.500 \\ 4.000 \\ 3.378 \\ 3.667 \\ 3.714 \\ 0.000 \end{pmatrix}, \; \bar{\upsilon}^{\text{int}} = \begin{pmatrix} 0.500 \\ 5.465 \\ 2.443 \\ 4.329 \\ 7.023 \\ 0.000 \end{pmatrix}. \quad (2.94)$$

A couple of remarks are in order. But first, to help clarify the discussion, let the nodes belonging to different components of the bow-tie topology be identified accordingly. The set of integer subscripts $\{\mathcal{I}\}$ denotes the IN-nodes. Of the set of IN-nodes, $\{\mathcal{R}\} \subset \{\mathcal{I}\}$ labels the actual root nodes. Nodes in the OUT section have indices $\{\mathcal{O}\}$. Moreover, $\{\mathcal{I}\} \cap \{\mathcal{O}\} = \emptyset$. The subscripts $\{\mathcal{S}\} := \{\mathcal{S}; \mathcal{S} \neq \mathcal{I} \wedge \mathcal{S} \neq \mathcal{O}\}$ denote the remaining nodes in the SCC (ignoring the T&T). The network value and integrated value can thus be symbolically dissected into these components. As an example, $\upsilon^{\text{net}} = \left(\upsilon^{\text{net}}_{\{\mathcal{I}\}}, \upsilon^{\text{net}}_{\{}\mathcal{S}\}, \upsilon^{\text{net}}_{\{\mathcal{O}\}}\right)^t$.

In the example seen in Fig. 2.4, $\{\mathcal{R}\} = \{\mathcal{I}\} = \{1\}$, $\{\mathcal{S}\} = \{2, 3, 4, 5\}$, and $\{\mathcal{O}\} = \{6\}$.

### Root Nodes

The problem of root nodes accumulating the sum of the underlying value of downstream nodes as network vale, e.g., $\tilde{\upsilon}^{\text{int}}_{\mathcal{R}} = \hat{\upsilon}^{\text{int}}_{\mathcal{R}} = 5.0$, disappears for $\mathring{\upsilon}^{\text{int}}_{\mathcal{R}} = \bar{\upsilon}^{\text{int}}_{\mathcal{R}} = 0.5$. This concludes that the analytical and algorithmic solutions presented in this thesis remedy the lamented problem.

Observe also, that for a single root node the corresponding entry in the correction operator is one. Hence $[\mathcal{D}\upsilon]_{\mathcal{R}} = \upsilon_{\mathcal{R}}$. Recalling Eqs. (2.77)–(2.81), this means that $\upsilon^{\text{net}}_{\mathcal{R}} - \tilde{\upsilon}^{\text{int}}_{\mathcal{R}} = \hat{\upsilon}^{\text{net}}_{\mathcal{R}} - \hat{\upsilon}^{\text{int}}_{\mathcal{R}} = \mathring{\upsilon}^{\text{net}}_{\mathcal{R}} - \mathring{\upsilon}\upsilon^{\text{int}}_{\mathcal{R}} = \bar{\upsilon}^{\text{net}}_{\mathcal{R}} - \bar{\upsilon}^{\text{int}}_{\mathcal{R}} = [\mathcal{D}\upsilon]_{\mathcal{R}} = \upsilon_{\mathcal{R}}$.

### Cycles

For the nodes in the SCC, the large values of $\upsilon^{\text{net}}_{\{\mathcal{S}\}}$ are decreased to $\hat{\upsilon}^{\text{net}}_{\{\mathcal{S}\}} = \mathring{\upsilon}^{\text{net}}_{\{\mathcal{S}\}}$ and $\bar{\upsilon}^{\text{net}}_{\{\mathcal{S}\}}$. As proven in Appendix D.1, the algorithmic network value is equivalent to the corrected one for root nodes. The quantities in $\bar{\upsilon}^{\text{net}}_{\{\mathcal{S}\}}$ are novel corrected network values. Because there is no straight-forward interpretation but only an analytical definition, cf. Eq. (2.77), mathematical consistency alone justifies the existence of this variant of network value. In essence, the original network value $\upsilon^{\text{net}}$ can be seen to be progressively transformed into the fully corrected form given by $\bar{\upsilon}^{\text{net}}$, with $\hat{\upsilon}^{\text{net}}$ and $\mathring{\upsilon}^{\text{net}}$ being the intermediate steps.

Note that although the actual numerical value sizes of $\bar{v}^{\text{net}}_{\{S\}}$ are comparable to those of $\hat{v}^{\text{net}}_{\{S\}}$, the order of its elements preserves the original order given in $v^{\text{net}}_{\{S\}}$. Ordering these nodes by descending network value yields the labels 5, 2, 4, 3. This is not the case for $\hat{v}^{\text{net}}_{\{S\}}$, where the same ordering gives: 2, 5, 4, 3.

For the integrated value, the correspondence of the network value variants in the SCC, i.e.,[6]

$$\hat{v}^{\text{net}}_{\{S\}} = \hat{\nu}^{\text{int}}_{\{S\}} + \mathcal{D}|_{S}\, v_{S} \tag{2.95a}$$

$$= \mathring{\nu}^{\text{int}}_{\{S\}} + v_{S} = \mathring{v}^{\text{net}}_{\{S\}}, \tag{2.95b}$$

is not retained:

$$\hat{\nu}^{\text{int}}_{\{S\}} \neq \mathring{\nu}^{\text{int}}_{\{S\}}. \tag{2.96}$$

By employing the correction operator $\mathcal{D}$ in the computation of $\mathring{\nu}^{\text{int}}_{\{S\}}$, the relationship would be restored, as a simple rearrangement of Eq. (2.95) reveals

$$\mathring{v}^{\text{net}}_{\{S\}} - \mathcal{D}|_{S}\, v_{S} = \hat{\nu}^{\text{int}}_{\{S\}}. \tag{2.97}$$

However, for the BFS algorithm, computing $\mathcal{D}$ from Eq. (2.51) would restrict the method's applicability to very large networks, as the inverse of the matrix $(I - W)$ is again the size of the whole network. This makes Eq. (2.97) unsuitable for the definition of integrated value in the BFS algorithm and we are left with the discrepancy[7]: $\hat{\nu}^{\text{int}}_{\{S\}} - \mathring{\nu}^{\text{int}}_{\{S\}}$. But overall, as the removal of incoming links to a node $i$ (when constructing the subnetwork of nodes downstream, as the BFS algorithm does) only introduces a deviation to the integrated value and not the network value, this is a minor problem in any case. Especially as the maximal difference between $\hat{\nu}^{\text{int}}$ and $\mathring{\nu}^{\text{int}}$ is bounded. This can been seen as follows: by construction, it is true that

$$\mathring{\nu}^{\text{int}} = \hat{\nu}^{\text{int}} - \text{diag}(\widehat{W})v, \tag{2.98}$$

or in scalar notation

$$\mathring{\nu}^{\text{int}}_{i} = \hat{\nu}^{\text{int}}_{i} - \widehat{W}_{ii} v_{i} = \hat{\nu}^{\text{int}}_{i} - \frac{V_{ii} - 1}{V_{ii}} v_{i}, \tag{2.99}$$

recalling Eq. (2.34b). From Eqs. (2.33) and (2.16), noting that per definition $W_{ij} \in [0, 1]$, it follows that

$$V_{ii} = 1 + W_{ii} + [W^2]_{ii} + \cdots \geq 1. \tag{2.100}$$

---

[6] Recall Eqs. (2.59) and (2.88) or (2.87).

[7] It is not clear if this difference should be understood as an error in the computation or if simply $\mathring{\nu}^{\text{int}}$ is just another legitimate variant to the theme of integrated value.

Hence

$$\hat{\nu}_i^{\text{int}} - \mathring{\nu}_i^{\text{int}} = \left(1 - \frac{1}{V_{ii}}\right) v_i. \tag{2.101}$$

As $(1 - 1/V_{ii}) \in [0, 1]$, the maximal difference of $\mathring{\nu}_i^{\text{int}}$ and $\hat{\nu}_i^{\text{int}}$ is $v_i$, i.e., as big as or smaller than the underlying value of the node $i$ itself.

A more rigorous derivation yields the exact quantification of the difference in the SCC. From Eq. (2.95) it can be derived that

$$\hat{\nu}_{\{\mathcal{S}\}}^{\text{int}} - \mathring{\nu}_{\{\mathcal{S}\}}^{\text{int}} = v_{\mathcal{S}} - \mathcal{D}|_{\mathcal{S}}\, v_{\mathcal{S}}. \tag{2.102}$$

A final observation is that the difference in the duality relation of Eq. (2.75) is tightly related to the above mentioned quantity:

$$|\widehat{W} v| - |W \hat{v}^{\text{net}}| = |v| - |\mathcal{D} v|, \tag{2.103}$$

where $|v| = \sum_i v_i$ is the norm of a vector $v$. It is an interesting fact that the failing of the duality relation (due to the implementation of the correction for cycles) can be expressed solely using the intrinsic value and the correction operator. In the example above, $|v| - |\mathcal{D}v| = 3.557$. Moreover, the difference between $\hat{\nu}^{\text{int}}$ and $\mathring{\nu}^{\text{int}}$ is identical to this value for SCC-nodes. This can be seen as justification that $\mathring{\nu}^{\text{int}}$ has an existence in its own right, as it emerges due to the correction.

This concludes the discussion of network value and integrated value in ownership networks, i.e., the different notions of value that can be derived from shareholding relations and a proxy for size or value of firms. The next question to be answered, in the quest to unveil the distribution of economic power worldwide, is: how to compute control from ownership relations? But before the concept of control is introduced in Sect. 2.7, in the next section the methodology presented so far is recast in a different context. It is straightforward to move away from the economic motivation and interpretation driving the above methods towards a general framing valid for generic complex and real-world networks.

## 2.6 The General Setting: The Notion of Centrality in Networks

In this chapter, the motivation and interpretation for the methodology was primarily given from an economics context. Only in Sect. 2.4 we generalized the concepts to generic networks and discovered the close correspondence between the integrated value and the notion of a quantity flowing in the network. Here we add another complementary point of view coming from centrality measures aiming at identifying the most important nodes of a certain network configuration.

The notion of centrality has a long history in social science as a structural attribute of nodes in a network, that depends on their position in the network (Hubbell 1965; Bonacich 1972; Freeman 1978). Centrality refers to the extent to which a network is centered around a single node. In a star network for example, the central node has the highest centrality, and all other nodes have minimum centrality.

Centrality is a fundamental concept in network analysis (Borgatti and Everett 2006). Recently, there has been a lot of work on centrality in networks in physics and biology (Freeman 2008) next to economics (Schweitzer et al. 2009). Most of the attention has been devoted to the feedback-type centrality which is discussed in the following.

This notions of centrality is based on the idea that a node is more central the more central its neighbors are themselves. The idea leads to a set of equations which need to be solved simultaneously. In general, this type of centrality is also categorized to as eigenvector centrality. The motto "the importance of a node depends on the importance of the neighboring nodes" can be quantified as

$$c_i = \sum_j A_{ij} c_j, \tag{2.104}$$

where $A$ is the adjacency matrix of the graph and $c_i$ denotes the centrality score of node $i$. In matrix notation

$$c = Ac. \tag{2.105}$$

A solution can be found if the equation is understood as an eigenvector equation

$$\lambda c = Ac. \tag{2.106}$$

The *Bonacich eigenvector centrality* reinterprets Eq. (2.106) in terms of centrality (Bonacich 1972)

$$c_i^B := \frac{1}{\lambda} \sum_j A_{ij} c_j^B, \tag{2.107}$$

with the solution

$$c^B = (\lambda I - A)^{-1} e, \tag{2.108}$$

$e$ being a column vector of ones.

The *Hubbell index* is defined for weighted directed graphs (Hubbell 1965). The nodes can be thought to posses an intrinsic importance $c^0$, to which the importance from being connected to other nodes is added

$$c^H := Ac^H + c^0, \tag{2.109}$$

The solution is

$$c^H = (I - A)^{-1} c^0, \tag{2.110}$$

which exists if $I - A$ is invertible or equivalently, if there is no eigenvalue of $A$ equal to one, $\lambda_i(A) \neq 1 \forall i$.

*α-Centrality* introduced in Bonacich and Lloyd (2001) is defined as

$$c^\alpha := \alpha A x + c^0, \tag{2.111}$$

the vector $c^0$ again assigning an initial centrality value and the solution is given by

$$c^\alpha = (I - \alpha A)^{-1} c^0. \tag{2.112}$$

An additional variant of eigenvector centrality is the *$c(\alpha, \beta)$-Centrality* introduced in Bonacich (1987)

$$c_i(\alpha, \beta) := \sum_j (\alpha + \beta c_j) A_{ij}, \tag{2.113}$$

with the solution

$$c(\alpha, \beta) = \alpha (I - \beta A)^{-1} A e, \tag{2.114}$$

$e$ being the column vector of ones.

Comparing $c^H$ and $c^{\alpha=1}$ with $v^{\text{net}}$ of Eq. (2.47) uncovers a close similarity of these measures. Moreover, setting $e = v$ in Eq. (2.114), reveals that $c(1, 1)$ corresponds to $\tilde{v}^{\text{int}}$ given in Eq. (2.42). Consequently and in general, network value and integrated value should be understood as centrality scores giving the importance of nodes in the network. $\tilde{v}^{\text{int}}$ considers the centrality of the neighbors, while $v^{\text{net}}$ employs $\tilde{v}^{\text{int}}$ plus an intrinsic centrality of the nodes themselves. It should be highlighted that the reinterpretation of network value and network control in terms of flow and centrality of generic networks is a novel contribution of this thesis.

Furthermore, we propose $\bar{v}^{\text{net}}$ and $\bar{v}^{\text{int}}$, cf. Eqs. (2.76) and (2.81), as new centrality measures for networks with bow-tie topologies. These novel quantities correct for self-cycles exaggerating the values and also solve the associated problem of root nodes accumulation. In a nutshell, $\tilde{W}$ should be replaced as

$$\tilde{W} = (I - W)^{-1} W \quad \longmapsto \quad \bar{W} = W \mathcal{D}(\tilde{W} + I), \tag{2.115}$$

as seen from Table 2.2.

Two final remarks. Firstly, Borgatti (2005) discusses the relationship between centrality and flow. Secondly, in Chap. 5 we present a network-evolution model based on different centrality measures. There we compare and discuss integrated value and network value and compare them with Google's Pagerank centrality, also discussed in Appendix B.7.

## 2.7 Moving from Ownership to Control

Until now, the methods discussed in this chapter where directly related to ownership networks. The different centrality measures seen in the last section are interpreted as a proxy for the economic value associated with a corporation entangled in a web of ownership relations. But what does all of this have to do with control? Or even more fundamental, what is control in this context?

Ownership is an objective quantity given by the percentage of shares owned in a company. In detail, these percentages of ownership in the equity capital of a firm, also referred to as cash-flow rights, are associated with so-called voting rights. Such votes assigned to the holders can be exercised at shareholder's meetings. The more voting rights a shareholder has in a corporation, the greater the influence that can be exerted over the company, thus the higher the level of control. See also Sect. 1.2.1. There is a great freedom in how corporations are allowed to map cash-flow rights into voting rights assigned to the shareholders (e.g., nonvoting shares, dual classes of shares, multiple voting rights, golden shares, voting-right ceilings, etc.). As a consequence, control can only be estimated. Several models aiming at deriving control based on the knowledge of ownership have been proposed. In this section we discuss these and in the next section introduce a novel model of control.

As in the case of ownership, the presence of the network should also affect the computation of control. In the previous sections, the notion of integrated ownership[8] was introduced in Eq. (2.15). A similar shift from direct to integrated control will conclude the methodology in Sect. 2.9, giving rise to the main theme of this thesis: the flow of control. In an intermediary step, at the end of this section, an alternative method for the propagation of control in a network is first discussed.

In essence, the models of control take the adjacency matrix of the ownership network and transform it into a matrix reflecting the control relations, or symbolically

$$W \longmapsto \mathcal{C}, \tag{2.116}$$

where $\mathcal{C}$ now depends on the chosen model of control.

### The "One-Share-One-Vote" Rule

Despite the mentioned freedom in how firms can issue voting rights to the shareholders on the basis of their cash-flow rights, empirical studies indicate that in many countries the corporations tend not to exploit all the opportunities allowed by national laws to skew voting rights. Instead, they adopt the so-called one-share-one-vote principle which states that ownership percentages yield identical percentages of voting rights (La Porta et al. 1999; The Deminor Group 2005; Goergen et al. 2005).

[8] Meaning on the basis of all direct and indirect paths in the network.

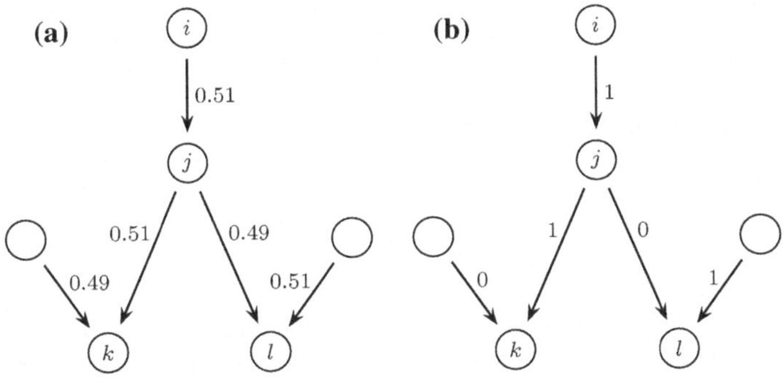


**Fig. 2.5** The threshold model (TM) is a simple way of estimating direct control from direct ownership; if the percentages of ownership seen in (**a**) exceed a certain threshold, taken to be 50 % here, the shareholder gets assigned full control (**b**)

According to this linear model (LM), there is no deviation between ownership and control, thus the direct control matrix coincides with the direct ownership matrix and Eq. (2.116) reveals the simple relation

$$\mathcal{C}_{ij}^{LM} := W_{ij}. \tag{2.117}$$

### The Threshold Model

The simple linear model of the last section overlooks one special trait of control, namely that it is often binary. As an example, holding over 50 % of the votes ensures that one has full or incontestable control. This feature is considered in the threshold model (TM), also referred to as majority model. Various values for the fixed threshold of absolute control have been proposed: 10–20 % (La Porta et al. 1999), next to a more conservative value of 50 % (Chapelle and Szafarz 2005).

For the control matrix $\mathcal{C}$ one finds

$$\mathcal{C}_{ij}^{TM} := \begin{cases} 1, \text{ if } W_{ij} > \vartheta, \\ 0, \text{ if } \exists k \neq i : W_{kj} > \vartheta, \\ W_{ij}, \text{ otherwise.} \end{cases} \tag{2.118}$$

where $\vartheta$ is the chosen threshold value. See Fig. 2.5 for an example illustrating the concept.

### The Control Value

The notion of the shareholders portfolio value was introduced in Eq. (1.2) of Sect. 1.1.3. It is a measure of the value gained from the direct ownership relations.

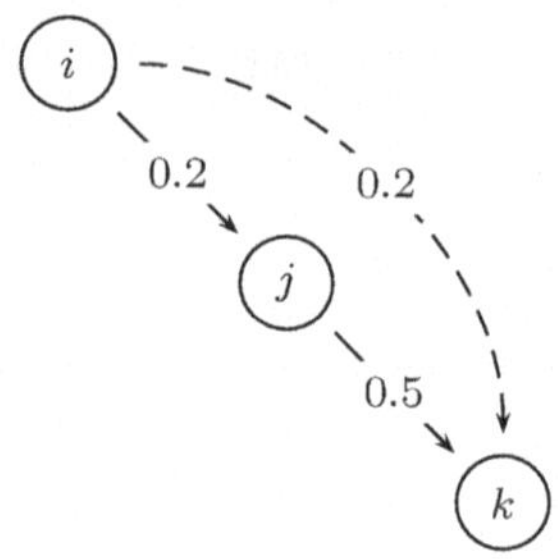


**Fig. 2.6** Application of the weakest link model: the control of company $i$ over company $k$ is given by the minimum percentage of control along the path of indirect control, i.e., 20 %

In a similar vein, this measure can be easily extended to reflect and incorporate the notions of control. By replacing $W_{ij}$ in Eq. (1.2) with a control matrix $\mathcal{C}_{ij}$, i.e., by symbolically applying Eq. (2.116), we introduce the *control value*:

$$c_i := \sum_{j=1}^{k_i^{out}} \mathcal{C}_{ij} v_j. \tag{2.119}$$

It is a measure of the economic value shareholder $i$ can control considering its direct ownership shares.

### Indirect Control

The two simple models presented above are examples of direct control estimations. However, how does control propagate along the links in a network?

The "weakest link" model proposes a measure of control which selects the weakest relation in a chain of control (Claessens and Djankov 2000). As seen in Fig. 2.6, if $i$ owns 20 % of the votes of firm $j$ and $j$ owns 50 % of the votes of $k$, the weakest link rule assigns to $i$ a control over $k$ of 20 %, the minimum between 20 % and 50 %. The intuition is given by the fact that an outsider can gain control of firm $k$ by acquiring a controlling stake of 20 % over firm $j$.

However, this methodology is not able to measure control in the case of complex ownership structures, such as cross-shareholdings. It is not clear how to adapt the method in light of the recursive nature of cross-shareholding relations. Moreover, for very long chains of indirect ownership, the weakest link model can overestimate the control the first firm in the chain has over the last one.

Another model for estimating control in networks is given by applying the integrated model to the control matrix $\mathcal{C}$. In the literature, however, the integrated model has nearly exclusively been applied to ownership adjacency matrices (Brioschi et al. 1989; Flath 1992; Baldone et al. 1998; Chapelle 2005). An exception being (Chapelle and Szafarz 2005), defining integrated control based on the TM. We can use Fig. 2.5 again as an example to highlight how this mechanism works. The shareholdings $W_{ij}$

get transformed into $C_{ij}^{TM}$ according to Eq. (2.118). Then, for instance, the indirect control form $i \to k$ is simply given by the multiplication $C_{ij}^{TM} C_{jk}^{TM}$, in analogy to the example seen in Fig. 2.1a on p. 24. In other words, $i$ has full control over $k$ via $j$. In general, this means that, for example by ignoring the problems of cycles,

$$\tilde{\mathcal{C}}^{TM} := (I - \mathcal{C}^{TM})^{-1} \mathcal{C}^{TM}, \tag{2.120}$$

employing Eq. (2.15). For the LM, the procedure is equivalent.

The notion of integrated control, in analogy to the integrated value defined in Eq. (2.42), is detailed in Sect. 2.9, after the introduction of of a new model of control in the next section.

## 2.8 The Relative Majority Model of Control

In the following, we introduce a new model for estimating control from ownership relations, extending the list of existing ones presented in the last section.

There is a very general problem plaguing the two models described in Sect. 2.7. Namely, that shareholders do not only act as individuals but can collaborate in shareholding coalitions that give rise to so-called voting blocks. The theory of political voting games in cooperative game theory has been applied to the problem of shareholder voting. There are four proposed ways to measure control in a relative manner.

The *fixed rule* simply classifies the degrees of control according to fixed thresholds of the leading shareholdings (Leech and Leahy 1991).

The *Herfindahl index*, or H-index, was originally used in economics as a standard measure of market concentration (Herfindahl 1959)

$$\mathcal{H} := \sum_i w_i^2, \tag{2.121}$$

where $w_i$ are some sort of market shares. It has been employed as a measure of how concentrated or dispersed ownership is (where $w_i$ are now the shareholdings of a specific firm) (Cubbin and Leech 1983; Demsetz and Lehn 1985; Leech 1988; Leech and Leahy 1991).

The so-called *power indices* were originally introduced as the Shapley-Shubik index (SS-index) by Shapley and Shubik (1954) and famously extended by Banzhaf (1965) to the Banzhaf index (B-index). They come from the notion of a weighted majority in cooperative games and measure the extent to which shareholders are pivotal to the success in potential voting pacts. Power indices measure the relative influence of shareholder over decision making. The SS-index considers orderings of $N$ players (permutations) while the B-index counts coalitions. Both give essentially similar results and use a majority rule (Prigge 2007). They provide a continuous variable which is connected to the share in votes in a non-linear manner. Selected empirical studies were done by Leech (1988), Crama et al. (2003), Chen (2004), Prigge (2007).

However, the employment of these game theoretic power indices for measuring shareholder voting behavior has failed to find widespread acceptance in the corporate finance literature for estimating control. Mainly due to computational, inconsistency and conceptual issues. First of all, there is an ambiguity with the definition of "power" (Prigge 2007). Secondly, when voters have varying weights, the results of the main two power indices and their variants all yield different results (Leech 2002a,b). Thirdly, the stock of empirical studies is rather small, and the few results are inconclusive (Prigge 2007). Fourthly, for a large number of agents, the computational demands become challenging (Leech 2002a,b). And finally, the notion of integrated ownership is also ignored.

The so-called *degree of control*, or $\alpha$, was introduced as a probabilistic voting model measuring the degree of control of a block of large shareholdings as the probability of it attracting majority support in a voting game (Cubbin and Leech 1983; Leech 1987a,b; Leech and Leahy 1991). Without going into details, the idea is as follows. Consider a shareholder $i$ with ownership $W_{ij}$ in the company $j$. Then the control of $i$ depends not only on the value in absolute terms of $W_{ij}$, but also on how dispersed the remaining shares are (measured by the Herfindahl index). The more they tend to be dispersed, the higher the value of $\alpha$. So even a shareholder with a small $W_{ij}$ can obtain a high degree of control. The assumptions underlying this probabilistic voting model correspond to those behind the power indices (Cubbin and Leech 1983; Leech 1987a,b; Leech and Leahy 1991; Chen 2004). It relates to the SS-index by treating all permutations as equally likely and to the B-index by treating all coalitions as equiprobable (Leech 1987b). The degree of control is closely related to a measure of a priori voting power defined for weighted voting games (Leech 1988).

However, $\alpha$ suffers from drawbacks. It gives a minimum cutoff value of 0.5 (even for arbitrarily small shareholdings, see also Appendix D.2) and hence Eq. (1.1) is violated, meaning that it cannot be utilized in an integrated model. The computation of $\alpha$ can become intractable in situations with many shareholders.

Having listed these issues, we present a minimal list of requirements a reasonable model of control should fulfill:

1. Define a mapping from $F : (0, 1]^N \rightarrow (0, 1]^N$, for the $N$ shareholding relations $\{W_{ij}\}$, where $F_1(\{W_{ij}\}), \ldots, F_N(\{W_{ij}\})$ represent control and take on continuous values.
2. Be extendable to an integrated model.
3. Sum to one for each firm, as $\sum_j W_{ij}$ in principle does.
4. Emulate the behavior of $\alpha$ for large shareholders (coalitions and voting blocks).
5. Have an intuitive meaning of controlling power.
6. Be feasible to compute on large networks.

In the following section, we introduce our new model.

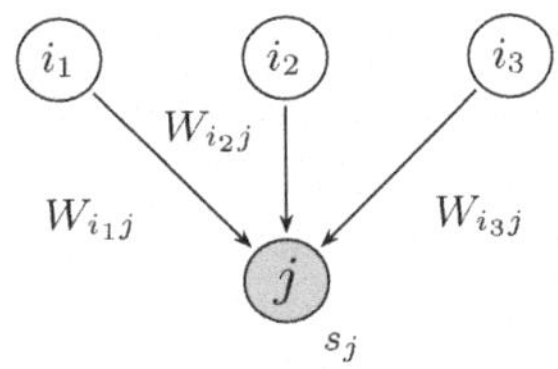


**Fig. 2.7** Definition of the concentration index $s_j$, measuring the number of prominent incoming edges, respectively the effective number of shareholders of the company $j$; when all the weights are equal, then $s_j = k_j^{in}$, where $k_j^{in}$ is the in-degree of vertex $j$; when one weight is overwhelmingly larger than the others, the concentration index approaches the value one, meaning that there exists a single dominant shareholder of $j$

### *2.8.1 Extending the Notions of Degree for Weighted Networks*

Although the concepts about to be introduced here were motivated as being related to the separation of ownership and control in economics, they are best understood in the context of pure network theory.

In this paradigm, we substantiate the idea of the 3-Level network analysis mentioned in Sect. 1.1. To recall, complex real-world networks can be understood at three levels of resolution: the topological, with weighted and directed links, and by assigning non-topological sate variables to the nodes.

The following measures can be understood as Level 2 quantities, extending previous Level 1 network notions. Namely, the degree and strength, explained in Appendix B.2.

These quantities are used in Chap. 3 for the extraction of the backbone (see more in Sect. 3.3.2). Some empirical distributions are given in Figs. 3.2 and 3.3 of Sect. 3.3.3.

#### In-Degree

When there are no weights associated with the edges, we expect all edges to count the same. If weights have a large variance, some edges will be more important than others. A way of measuring the number of prominent incoming edges is to define the *concentration index* (Battiston 2004) as follows:

$$s_j := \frac{\left(\sum_{i=1}^{k_j^{in}} W_{ij}\right)^2}{\sum_{i=1}^{k_j^{in}} W_{ij}^2}. \tag{2.122}$$

If the equality in Eq. (1.1) holds, the numerator will be equal to one. Observe that this quantity is akin to the inverse of the Herfindahl index of Eq. (2.121). Notably, a similar measure has also been used in statistical physics as an order parameter (Derrida and Flyvbjerg 1986). A recent study (Serrano et al. 2009) employs a Herfindahl index in their backbone extraction method for weighted directed networks (where, however, the nodes hold no non-topological information). In the context of ownership networks, $s_j$ is interpreted as the effective number of shareholders of the firm $j$, as explained in Fig. 2.7. Thus it can be understood as a measure of control from the point of view of a company.

#### Out-Degree

The second quantity to be introduced measures the number of important outgoing edges of the vertices. For a given vertex $i$, with a destination vertex $j$, we first define a measure which reflects the importance of $i$ with respect to all vertices connecting to $j$:

$$H_{ij} := \frac{W_{ij}^2}{\sum_{l=1}^{k_j^{in}} W_{lj}^2}. \tag{2.123}$$

This quantity has values in the interval (0, 1]. For instance, if $H_{ij} \approx 1$ then $i$ is by far the most important source vertex for the vertex $j$. For our ownership network, $H_{ij}$ represents the *fraction of control* (Battiston 2004) shareholder $i$ has on the company $j$. As shown in Fig. 2.8, this quantity is a way of measuring how important the outgoing edges of a node $i$ are with respect to its neighbors' neighbors. For an interpretation of $H_{ij}$ from an economics point of view, consult the following section.

From this, we then define the *control index*:

$$h_i := \sum_{j=1}^{k_i^{out}} H_{ij}. \tag{2.124}$$

Within the ownership network setting, $h_i$ is interpreted as the effective number of stocks controlled by shareholder $i$. In essence, $s$ and $h$ replace the in- and out-degree in the case of weighted and directed networks.

### *2.8.2 Interpretation as a New Model of Control*

The definition of the fraction of control $H_{ij}$ given in Eq. (2.123), can be understood as yielding a new non-linear control model, that lies between the linear mapping given by the LM, cf. Eq. (2.117), and the digital threshold-driven TM, seen in Eq. (2.118). As this model assigns control based on the relative fraction of ownership shares that each shareholder has, it is called the *relative majority model of control*, or simply the

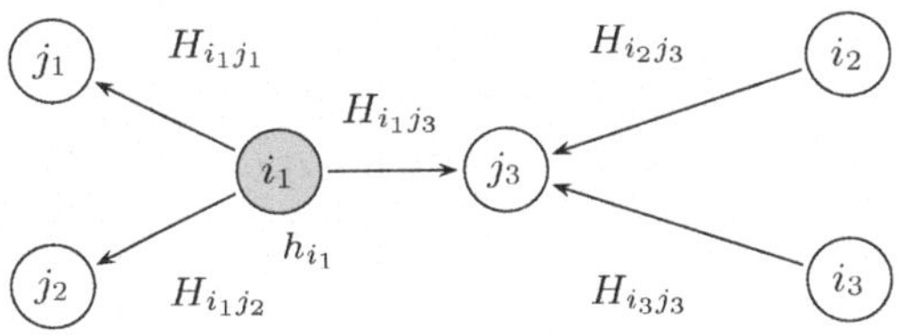


**Fig. 2.8** The definition of the control index $h_i$, measuring the number of prominent outgoing edges; in the context of ownership networks this value represents the effective number of firms that are controlled by shareholder $i$; note that to obtain such a measure, we have to consider the fraction of control $H_{ij}$, which is a model of how ownership can be mapped to control (see the discussion in Sect. 2.8.2)

relative model (RM). In other words

$$c_{ij}^{RM} := H_{ij}. \tag{2.125}$$

In summary, our quantity $H_{ij}$ adheres to the small catalogue of desired features presented in the list on p. 56. It holds that $\sum_j H_{ij} = 1$, for all firms $j$. In effect, any shareholder gaining control will be offset by shareholders loosing control. As a result, this measure of control can also be used as an integrated model, by applying Eq. (2.15) to yield $\widetilde{H}_{ij}$. For large shareholders, the analytical expressions of $H_{ij}$ and $\alpha$ share very similar behavior, as detailed in Appendix D.2. This means that to some extent our measure of control can take possible strategic alliances of shareholders into account without requiring the knowledge of data on voting blocks. There is an intuitive meaning of power associated with our model: how important is a shareholder with respect to all other shareholders, or what is the relative voting power of a shareholder considering the dispersion of the rest of the votes? We are able to compute $\widetilde{H}_{ij}$ for every shareholder in the sample without facing any computational restrictions. To summarize, the properties of our model make a sensible ranking of all shareholders according to their controlling power possible.

This concludes that the new measure of control merges crucial insights from the corporate finance literature and the game theoretic approach to voting while addressing their mentioned shortcomings. It should also be noted, that $s_j$ represents the complementary of $h_i$: while the latter represents the control seen from the point of view of the shareholders, the former reflects the control seen by the firms.

## 2.9 Computing the Flow of Control in an Ownership Network

In this section we conclude the methodology chapter. To summarize, the existing notion of integrated ownership, considering all direct and indirect paths of ownership, was generalized for generic ownership networks: $\widetilde{W}$, given in Eq. (2.15). This allowed the introduction of the integrated (portfolio) value $\tilde{\upsilon}^{\text{int}}$, defined in Eq. (2.42). The associated notion of network value $\upsilon^{\text{net}}$ reflects the integrated value of a firm plus its own intrinsic or underlying value, as seen in Eq. (2.49). These measures capture how the value of firms flow in a network of ownership relations, as described in Sect. 2.4.1.

It was shown in Sect. 2.2.4 that $\tilde{\upsilon}^{\text{int}}$ and $\upsilon^{\text{net}}$ faced problems when applied to networks with a bow-tie structure, because of the presence of cross-shareholding relations in the SCC. In the literature $\widehat{W}$ was proposed as a remedy, seen in Eq. (2.32). We fully implemented this correction by introducing $\hat{\upsilon}^{\text{int}}$ and $\hat{\upsilon}^{\text{net}}$, given in Eqs. (2.56) and (2.59).

In a next step, an additional problem of these novel measures was identified, see Sect. 2.3.5. In Sect. 2.5 we proposed two solutions, an algorithmic and an analytical one. As a result, $\bar{\upsilon}^{\text{int}}$ and $\bar{\upsilon}^{\text{net}}$ were defined in Eqs. (2.76) and (2.81). As well as $\mathring{\upsilon}^{\text{int}}$ and $\mathring{\upsilon}^{\text{net}}$ given in Eqs. (2.90) and (2.89).

The context given by the economic nature of the concepts was extended by noting the relation to centrality measures in networks, as described in Sect. 2.6. And returning to an economics setting, the notion of control, which can be derived from the knowledge of the ownership relations, was introduced in Sect. 2.7. This resulted in the two definitions of the matrix of control $C$, depending on the chosen model of control, seen in Eqs. (2.117) and (2.118).

We introduced a new model of relative direct control incorporating ideas originating in game theory in Sect. 2.8. It was noted how this new measure, defined in Eq. (2.123), can also be viewed as a pure network theoretic quantity extending the idea of degree for weighted and directed networks. In other words, it is a Level 2 measure.

Putting everything together, we arrive at a way to estimate corporate control in ownership networks. Parts of these methods were first published in Glattfelder and Battiston (2009) and Vitali et al. (2011).

Let $C$ be a control matrix based on one of the three models of direct control: LM, TM and RM. I.e., $C \in \{C^{LM}, C^{TM}, C^{RM}\} = \{W, C^{TM}, H\}$, recalling Eqs. (2.117), (2.118) and (2.125), respectively (2.123).

From this one can define *integrated control*, in analogy to integrated value, as

$$\dot{\varsigma}^{\text{int}} := \dot{C}\upsilon, \tag{2.126}$$

with $\upsilon_i$ being the intrinsic value of the firm $i$ and the symbol "˙" acting as a placeholder for "~", "^" or "¯", i.e., the chosen integrated model. Recall Table 2.2 for a summary of the corresponding definitions. In other words, integrated control measures the economic value a shareholder can control taking into account the network of firms in which it has direct and indirect shares (a Level 2 quantity). In addition, this last piece of the puzzle is in fact also a true Level 3 network measure. This means that it incorporates all the available information of the complex network under study: the weights and direction of links and (a proxy of) the intrinsic value or size firms, the non-topological state variable.

Finally, network value finds its correspondence in the so-called *network control*, defined as

$$c^{\text{net}} := \tilde{\varsigma}^{\text{int}} + v, \tag{2.127}$$

$$\hat{c}^{\text{net}} := \hat{\varsigma}^{\text{int}} + \mathcal{D}v, \tag{2.128}$$

$$\bar{c}^{\text{net}} := \bar{\varsigma}^{\text{int}} + \mathcal{D}v. \tag{2.129}$$

The network control of an economic actor is given by its intrinsic value plus the controlled value gained from the integrated control. For the algorithm described in Sect. 2.5.2, resulting in the integrated value and network value introduced in Eqs. (2.89) and (2.90) of Sect. 2.5.3, the corresponding analytical measures for control are

$$\overset{\circ}{\varsigma}{}^{\text{int}} = \overset{\circ}{c}{}^{\text{net}} - v, \tag{2.130}$$

where the algorithm computes $\overset{\circ}{c}{}^{\text{net}}$. The real-world meaning of these measures is discussed in Sect. 6.1, especially Sects. 6.1.8 and 6.1.3 offering an interpretation of integrated control in terms of potential power.

The knowledge of network control or integrated control can answer two questions:

1. Who are the most important economic actors in terms of control, ranked in descending order?
2. How is control distributed in the network?

In order to tackle the second question, it is necessary to find a way to measure the concentration of control. Such a concept is introduced in the following section.

## 2.10 Measuring the Concentration of Control

One last method needs to be introduced in order to round off this chapter. It is a general procedure for which the concentration of a random variable $X > 0$, drawn for all members of a given population, can be assessed. Figure 2.9 shows some possible examples of the distribution of $X$.

The methodology is similar to the construction of the Lorenz curve, uncovering the distribution of value in a market. In economics, the Lorenz curve gives a graphical representation of the cumulative distribution function (CDF) of a probability distribution. It is often used to represent income distributions, where the $x$-axis ranks the poorest $x$ % of households and relates them to a percentage value of income on the $y$-axis.

In our version, we invert the ordering on the $x$-axis and rank the shareholders according to their importance, as measured by network or integrated control, and report the fraction they represent with respect to the whole set of shareholder. The $y$-axis shows the corresponding percentage of controlled market value. In detail, we relate the fraction of shareholders to the fraction of the total value they collectively represent.

In generic terms, the population, taken to be comprised of $N$ individuals, is sorted by decreasing $X_i$ values. Without loss of generality, the individuals are labelled with increasing indices. The total amount distributed in this population is given by

$$X_{tot} := \sum_{i=1}^{N} X_i. \tag{2.131}$$

The individual with the highest value of the random variable has $X_1/X_{tot}$ percent of the total and represents $1/N$ percentage of the population. This corresponds to the first data point in the lower left-hand corner of the plots in Fig. 2.10. Similarly, the top right-hand corner of the diagrams represent 100 % of the population making up 100 % of the total. The concentration of $X$ is thus defined as the set of data points $(\eta, \vartheta)$, with

$$\eta(n) := \frac{n}{N}, \tag{2.132}$$

and

$$\vartheta(n) := \frac{1}{X_{tot}} \sum_{i=1}^{n} X_n, \tag{2.133}$$

$n \in [1, N]$.
In order to understand what this method reveals, we compare some general probability distributions and the level of concentration they are associated with. We choose the following families of probability density functions (PDF)

$$\mathcal{P}_{\text{sl}}(X; C, \alpha) := CX^{-\alpha}, \tag{2.134}$$

$$\mathcal{P}_{\text{exp}}(X; \lambda) := \lambda e^{-\lambda X}, \tag{2.135}$$

$$\mathcal{P}_{\text{ln}}(X; \mu, \sigma) := \frac{1}{X\sigma\sqrt{2\pi}} e^{-\frac{(\ln X - \mu)^2}{2\sigma^2}}. \tag{2.136}$$

In other words, $P_{\text{sl}}$, $P_{\text{ln}}$ and $P_{\text{exp}}$ describe scaling-law, log-normal and exponential distributions, respectively.

In Fig. 2.9, three scaling laws ($\alpha = 0.7, \alpha = 1.0, \alpha = 1.5$), one exponential ($\lambda = 0.02$) and two log-normal distributions ($(\mu, \sigma) = (3, 0.6), (\mu, \sigma) = (9.1, 0.2.4)$)

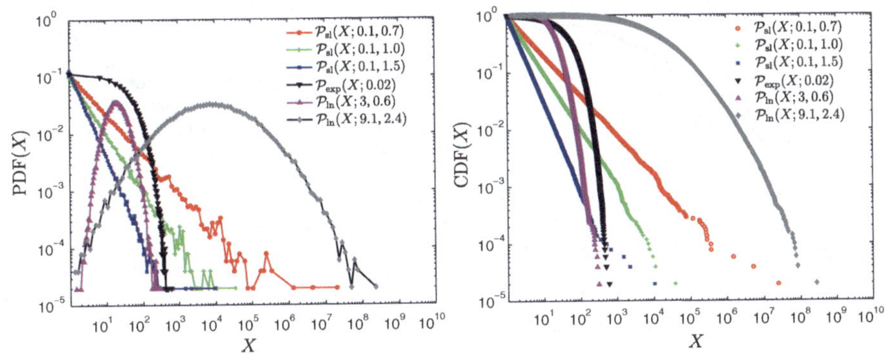


**Fig. 2.9** Plotting the PDFs (*left*) and CDFs (*right*) of the probability distributions given in Eqs. (2.134)–(2.136) for a random variable $X$; the plots are in log-log scale

are shown. The corresponding concentration is seen in Fig. 2.10. The semi-log scale representation in the bottom panel reveals the clearest picture. The scaling-law distribution with $\alpha = 0.7$ yields the highest concentration. The most important individual has a very large fraction of over 60 % of the total. It is an interesting observation, that the two remaining scaling-law distributions result in much lower concentration. Surprisingly, a log-normal distribution with a wide range of $X_i$, as given by $(\mu, \sigma) = (9.1, 0.2.4)$, is more concentrated than the scaling law with $\alpha = 1.5$ for nearly the whole range.

To summarize, the measure of concentration we propose is not only sensitive to the tail of the probability distribution, but also the relative distribution of mid-range values matters.

The cross-country analysis of Chap. 3 employs a variant of the above described procedure, called cumulative control. The diagram is discussed in Sect. 3.4.1. In the empirical analysis of Chap. 4, Sect. 4.3.3 shows the control distribution for the global network of corporations.

## 2.11 A Brief Summary

In the existing literature, only the notions of integrated ownership ($\widetilde{W}$ and $\widehat{W}$) and network (or group) value $\upsilon^{\text{net}}$ were introduced. The novel ideas presented in this thesis are the following. From an economics perspective:

1. the introduction of integrated value or control ($\tilde{\upsilon}^{\text{int}}$, $\hat{\upsilon}^{\text{int}}$, $\bar{\upsilon}^{\text{int}}$, $\tilde{\varsigma}^{\text{int}}$, $\hat{\varsigma}^{\text{int}}$, $\bar{\varsigma}^{\text{int}}$, $\mathring{\varsigma}^{\text{int}}$);
2. the idea of network control $c^{\text{net}}$;
3. the corrected network value and control ($\hat{\upsilon}^{\text{net}}$, $\bar{\upsilon}^{\text{net}}$, $\hat{c}^{\text{net}}$, $\bar{c}^{\text{net}}$, $\mathring{c}^{\text{net}}$);
4. the relative model of control $H_{ij}$;

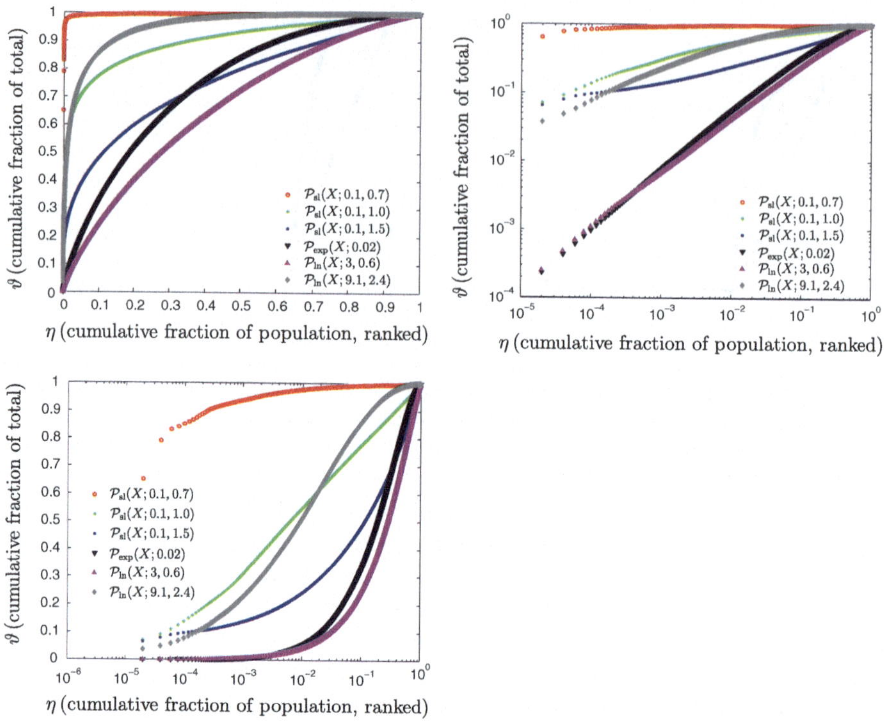


**Fig. 2.10** Concentration of the random variable $X$ resulting from the probability distributions shown in Fig. 2.9; the three plots are all identical, but displayed with different scales: (*top left panel*) linear plot, (*top right panel*) log-log plot and (*bottom panel*) semi-log plot; the construction of these curves is similar to the Lorenz curve used in economics, as described in the main text

5. the method to measure the concentration of a random variable;
6. the connection between network value or control, integrated value or control and the underlying value, e.g, $v^{\text{net}} = \tilde{\nu}^{\text{int}} + v$;
7. the identification of the correction operator $\mathcal{D}$.

From a complex-networks perspective:

1. the generalization of the methodology in terms of flow;
2. the interpretation of the methods as centrality measures;
3. the explanation of integrated value or control as true Level 3 quantities, incorporating the weights and direction of links next to non-topological state variables.

Having set aside all the required tools in order to estimate the flow of control in ownership networks, in Chaps. 3 and 4, these methods will be applied in two different empirical studies. The analysis uncovers important patterns and unveils the structural organization of ownership networks.

# References


H. Almeida, S.Y. Park, M. Subrahmanyam, D. Wolfenzon, Beyond cash flow and voting rights: valuation and performance of firms in complex ownership structures. Working paper, New York University Stern School of Business, 2007

S. Baldone, F. Brioschi, S. Paleari, Ownership measures among firms connected by cross-shareholdings and a further analogy with input-output theory, in 4th JAFEE international conference on investment and derivatives, 1998

J.F. Banzhaf, Weighted voting doesn't work: a mathematical analysis. Rutgers Law Rev. **19**, 317–343 (1965)

S. Battiston, Inner structure of capital control networks. Physica A **338**(1–2), 107–112 (2004)

P. Bonacich, Factoring and weighting approaches to status scores and clique identification. J. Math. Sociol. **2**(1), 113–120 (1972)

P. Bonacich, Power and centrality: a family of measures. Am. J. Sociol. **92**(5), 1170–1182 (1987)

P. Bonacich, P. Lloyd, Eigenvector-like measures of centrality for asymmetric relations. Soc. Netw. **23**(3), 191–201 (2001)

S.P. Borgatti, Centrality and network flow. Soc. Netw. **27**(1), 55–71 (2005)

S.P. Borgatti, R. Everett, A graph-theoretic perspective on centrality. Soc. Netw. **28**(4), 466–484 (2006)

F. Brioschi, S. Paleari, How much equity capital did the Tokyo stock exchange really raise. Finan. Eng. Japan. Mark. **2**(3), 233–258 (1995)

F. Brioschi, L. Buzzacchi, M.G. Colombo, Risk capital financing and the separation of ownership and control in business groups. J. Bank. Finance **13**(1), 747–772 (1989)

A. Chapelle, Separation of ownership and control: where do we stand? Corp. Ownersh. Control **2**(2), 91–101 (2005)

A. Chapelle, A. Szafarz, Controlling firms through the majority voting rule. Physica A **355**, 509–529 (2005)

Y.H. Chen, Voting power, control rents and corporate governance: an integrated analysis of owner control and corporate governance. *SSRN eLibrary*, 2004

S. Claessens, S. Djankov, The separation of ownership and control in East Asian corporations. J. Finan. Econ. **58**(1–2), 81–112 (2000)

Y. Crama, L. Leruth, L.D.R. Renneboog, Corporate control concentration measurement and firm performance. Discussion paper, Tilburg University, 2003. http://ideas.repec.org/p/dgr/kubtil/200317.html

J. Cubbin, D. Leech, The effect of shareholding dispersion on the degree of control in British companies: theory and measurement. Econ. J. **93**(370), 351–369 (1983)

H. Demsetz, K. Lehn, The structure of corporate ownership: causes and consequences. J. Polit. Econ. **93**(6), 1155–1177 (1985)

B. Derrida, H. Flyvbjerg, Multivalley structure in Kauffman's model: analogy with spin glasses. J. Phys. A.**19**, 1003–1008 (1986)

D. Flath, Indirect shareholding within Japan's business groups. Econ. Lett. **38**(2), 223–227 (1992)

L.C. Freeman, Centrality in social networks conceptual clarification. Soc. Netw. **1**, 215–239 (1978)

L.C. Freeman, Going the wrong way on a one-way street: centrality in physics and biology. J. Soc. Struct. **9**(2), (2008)

J.B. Glattfelder, S. Battiston, Backbone of complex networks of corporations: the flow of control. Phys. Rev. E **80**(3), 36104 (2009)

M. Goergen, M. Martynova, L. Renneboog, Corporate governance convergence: evidence from takeover regulation reforms in Europe. Oxford Rev. Econ. Policy **21**(2), 243–268 (2005)

O.C. Herfindahl, *Copper Costs and Prices: 1870–1957* (John Hopkins University Press, Baltimore, 1959)

C.H. Hubbell, An input-output approach to clique identification. Sociometry **28**(4), 377–399 (1965)

R. La Porta, F. Lopez de Silanes, A. Shleifer, Corporate ownership around the world. J. Finan. **54**(2), 471–517 (1999)

D. Leech, Corporate ownership and control: a new look at the evidence of Berle and Means. Oxford Econ. Pap. **39**(3), 534–551 (1987a)

D. Leech, Ownership concentration and the theory of the firm: a simple-game-theoretic approach. J. Ind. Econ. **35**(3), 225–240 (1987b)

D. Leech, The relationship between shareholding concentration and shareholder voting power in British companies: a study of the application of power indices for simple games. Manage. Sci. **34**(4), 509–527 (1988)

D. Leech, Computation of power indices. The Warwick Economics Research Paper Series (TWERPS) 644, University of Warwick, Department of Economics, 2002a. URL http://ideas.repec.org/p/wrk/warwec/644.html

D. Leech, An empirical comparison of the performance of classical power indices. Polit. Stud. **50**(1), 1–22 (2002b)

D. Leech, J. Leahy, Ownership structure, control type classifications and the performance of large British companies. Econ. J. **101**(409), 1418–1437 (1991)

W.W. Leontief, *Input-Output Economics* (Oxford University Press, Oxford, 1966)

S. Prigge, The performance of shareholder influence measures: an empirical assessment of share in votes, Banzhaf index, and Shapley-Shubik index. *SSRN eLibrary*, 2007

G. Rohwer, U. Pötter, *Transition Data Analysis User's Manual* (Ruhr-University, Bochum, 2005)

F. Schweitzer, G. Fagiolo, D. Sornette, F. Vega-Redondo, A. Vespignani, D.R. White, Economic networks: the new challenges. Science **325**(5939), 422 (2009)

M. Serrano, M. Boguñá, A. Vespignani, Extracting the multiscale backbone of complex weighted networks. Proc. Nat. Acad. Sci. **106**(16), 6483 (2009)

L.S. Shapley, M. Shubik, A method for evaluating the distribution of power in a committee system. Am. Polit. Sci. Rev. **48**(3), 787–792 (1954)

The Deminor Group. Application of the one share—one vote principle in Europe. Technical report, A report prepared for Association of British Insurers by Deminor, 2005. http://www.abi.org.uk/Bookshop/

S. Vitali, J.B. Glattfelder, S. Battiston, The network of global corporate control. PloS ONE **6**(10), e25995 (2011)

E. Wigner, The unreasonable effectiveness of mathematics in the natural sciences. Commun. Pure Appl. Math. **13**(1), 222–237 (1960)

# Chapter 3
# Backbone of Complex Networks of Corporations: The Flow of Control

*At each stage [of complexity] entirely new laws, concepts, and generalizations are necessary [...]. Psychology is not applied biology, nor is biology applied chemistry.*

(P.W. Anderson in Anderson 1972)

This chapter is based on the paper Glattfelder and Battiston (2009). The copyright of some of the images belongs to the *American Physical Society*. Note that in order to make the chapter self-consistent and self-supporting, some redundancies with Chaps. 1 and 2 are taken into account.

## 3.1 Introduction

In this chapter, we investigate the nature of ownership networks in 48 countries' stock markets. The issue of how control and wealth is structured is addressed in detail.

As a first step, we propose a new model to estimate corporate control based on the knowledge of the ownership ties. We then not only incorporate all three levels of network analysis (recall Sect. 1.1.1), but also consider higher orders of neighborhood relations, next to accounting for all indirect ownership ties in our study. In this respect, to our knowledge, there exists no comparable work of this kind in the literature. In fact when analyzing real-world networks, considering all three levels can yield new insights which would otherwise remain unobserved. For instance, in the current chapter, the identification of the key players in the networks under study is only possible if the network analysis takes into account a non-topological variable (namely, the value of the market capitalization of the listed companies).

However, considering all three levels of detail does not guarantee *per se* that new insights can be gained. It is also essential that the standard measures utilized in the analysis of complex networks are appropriately adapted to the specific nature of the network under investigation. For instance, the study of the degree distribution (details in Appendix B.2) in various real-world networks has revealed a universal feature across different domains: a scaling-law probability distribution. Such networks are

J. B. Glattfelder, *Decoding Complexity*, Springer Theses,
DOI: 10.1007/978-3-642-33424-5_3, © Springer-Verlag Berlin Heidelberg 2013

also called scale-free networks, see Appendix B.3. In many cases however, the degree of the nodes is not a suitable measure of connectivity (Barrat et al. 2004; Garlaschelli et al. 2005). In this paper, we introduce novel quantities, analogous to in-and out-degree, which are better suited for networks in which the relative weight of the links are important.

Our methodology allows us to identify and extract the core subnetwork where most of the value of the stock market resides, called the *backbone of control*. The analysis of these structures reveals previously unobservable results. Not only is the local dispersion of control accompanied by the concentration of control (and economic value) at the aggregate or global level, in addition, the local concentration of control is related to a global dispersion of control (and value). In detail, an even distribution of control at the level of individual corporations (typical of Anglo-Saxon markets) is accompanied by a high concentration of control and value at the global level. This novel observation means that, in such countries, although stocks tend to be held by many shareholders, the market as a whole is actually controlled by very few shareholders. On the other hand, in countries where the control is locally concentrated (e.g., European states), control and value is dispersed at the global level, meaning that there is a large number of shareholders controlling few corporations. Our empirical results are in contrast with previously held views in the economics literature (Davis 2008), where a local distribution of control was not suspected to systematically result in global concentration of control and vice versa. This emphasizes the fact that the bird's-eye-view given by a network perspective is important for unveiling overarching relationships.

Notably, we also provide a generalization of the method applicable to networks in which weights and direction of links, as well as non-topological state variables assigned to the nodes play a role. In particular, the method is relevant for networks in which there is a flow of mass (or energy) along the links and one is interested in identifying the subset of nodes where a given fraction of the mass of the system is flowing. The growing interest in methods for extracting the backbone of complex networks is witnessed by recent work in similar direction (Serrano et al. 2009).

The chapter is organized as follows. Section 3.2 describes the dataset we used. In Sect. 3.3 we introduce and discuss our methodology and perform a preliminary topological analysis of the networks. Section 3.4 describes the backbone extraction algorithm. In particular, we show that the method can be generalized by providing a recipe for generic weighted and directed networks. The section also introduces classification measures which are employed for the backbone analysis in Sect. 3.5. Finally, Sect. 3.6 summarizes our results and concludes this chapter.

## 3.2 The Dataset

We are able to employ a unique dataset consisting of financial information of listed companies in national stock markets. The ownership network is given by the web of shareholding relations from and to such companies, as depicted in Fig. 1.3. We constrain our analysis to a subset of 48 countries: United Arab Emirates (AE), Argentina (AR), Austria (AT), Australia (AU), Belgium (BE), Bermuda (BM), Canada (CA), Switzerland (CH), Chile (CL), China (CN), Germany (DE), Denmark (DK), Spain (ES), Finland (FI), France (FR), United Kingdom (GB), Greece (GR), Hong Kong (HK), Indonesia (ID), Ireland (IE), Israel (IL), India (IN), Iceland (IS), Italy (IT), Jordan (JO), Japan (JP), South Korea (KR), Kuwait (KW), Cayman Islands (KY), Luxembourg (LU), Mexico (MX), Malaysia (MY), Netherlands (NL), Norway (NO), New Zealand (NZ), Oman (OM), Philippines (PH), Portugal (PT), Saudi Arabia (SA), Sweden (SE), Singapore (SG), Thailand (TH), Tunisia (TN), Turkey (TR), Taiwan (TW), USA (US), Virgin Islands (VG), South Africa (ZA). In the following, the countries will be identified by their two letter ISO 3166-1 alpha-2 codes given in the parenthesis above. To assemble the ownership networks of the individual countries, we select the stocks in the country's market and all their available shareholders, who can be natural persons, national or international corporations themselves, or other legal entities.

The data is compiled from Bureau van Dijk's ORBIS database.[1] In total, we analyze 24877 corporations (or stocks) and 106141 shareholding entities who cannot be owned themselves (individuals, families, cooperative societies, registered associations, foundations, public authorities, etc.). Note that because the corporations can also appear as shareholders, the network does not display a bipartite structure. The stocks are connected through 545896 ownership ties to their shareholders. The database represents a snapshot of the ownership relations at the beginning of 2007. The values for the market capitalization, which is defined as the number of outstanding shares times the firm's market price, are also from early 2007. These values will be our proxy for the size of corporations and hence serve as the non-topological state variables.

We ensure that every node in the network is a distinct entity. In addition, as theoretically the sum of the shareholdings of a company should be 100 %, we normalize the ownership percentages if the sum is smaller due to unreported shareholdings. Such missing ownership data is nearly always due to their percentage values being very small and hence negligible.

## 3.3 A 3-Level Network Analysis

Not all networks can be associated with a notion of flow. For instance, in the international trade network the fact that country A exports to B and B exports to C, does not imply that goods are flowing from A to C. In contrast, in ownership networks the distance between two nodes (along a directed path) corresponds to a precise economic meaning which can be captured in a measure of control that considers all directed paths of all lengths (see Sect. 3.3.4). In addition, the weight of an ownership

[1] http://www.bvdep.com/orbis.html.

link has a meaning relative to the weight of the other links attached to the same node. Finally, the value of the nodes themselves is also very important.

Most network analysis focuses on topics like degree distribution, assortativity, clustering coefficients, average path lengths, connected components, etc. However, our specific interest in the structure of control renders most of these quantities inappropriate.

For instance, in an ownership network, the out-degree measures the number of firms in which a shareholder has invested. A high out-degree does not imply high control since the shares could be very small. Similarly, the in-degree, revealing the number of shareholders a corporation has, gives little insight into the amount of influence these shareholders can exert. In Sect. 3.3.2 we therefore extend the notion of degree to fit our context. Consequently, it is also not clear how to interpret degree-degree correlations, i.e., (dis-) assortativity.

The clustering coefficient defined for undirected graphs is equivalent to counting the number of triangles in a network. It does not have an obvious interpretation in the directed case, since an undirected triangle can correspond to several directed triangle configurations. Clustering coefficients have been introduced for weighted and undirected networks (Barrat et al. 2004), next to weighted and directed networks (Fagiolo 2007). However, these definitions only consider paths of length two. In contrast, in this paper, we use a measure of control that consider all paths of all lengths (see Sect. 2.9). Indeed, the knowledge of all the stocks reachable from any particular shareholder represents nothing else than a definition of indirect control.

For similar reasons, the average path length for the undirected graph does not have an interpretation in terms of control. Therefore, for our purposes, it also does not make sense to compute the small-world property (which is based on the two previously discussed quantities) of these real-world networks.

On the other hand, an analysis of the connected components may provide insights into the degree of fragmentation of the capital markets and we briefly address this issue in the following section. We then introduce extensions of existing network measures and define new quantities that better suit the ownership networks which are subsequently analyzed at all three levels of resolution in Sect. 3.4.

### *3.3.1 Level 1: Topological Analysis*

The network of ownership relations in a country is very intricate and a cross-country analysis of some basic properties of these networks reveals a great level of variability.

For example, an analysis of the number and sizes of connected components unveils a spectrum ranging from a single connected component in IS to 459 in the US. With

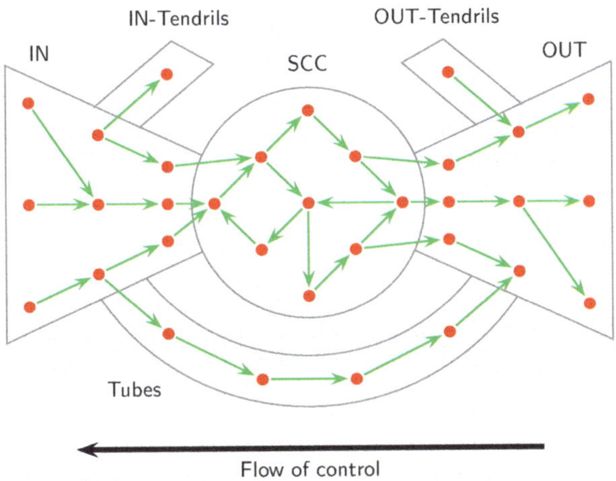


**Fig. 3.1** Illustration of a bow-tie topology: the central area is the strongly connected component (SCC), where there is a path from each node to every other node, and the *left* (IN) and *right* (OUT) sections contain the incoming and outgoing nodes, respectively; the *arrow* indicates how control flows in the network; as an example, subsidiaries would be located in the OUT, while natural persons are represented by leaf nodes in the IN; the SCC is comprised of firms connected by many cross-shareholding relations; in an ownership network, the direction of the links indicates the direction that the money, allowing an equity stake in a firm to be held, flows; in contrast, control is transferred in the opposite direction

a size of 18468, the largest connected component in the US is bigger than any single national ownership network in our sample.

Many small components correspond to a fragmented capital market while a giant and dense component corresponds to an integrated market. It is however not very clear what such connected components reveal about the structure and distribution of control. The same pattern of connected components can feature many different configurations of control. Therefore, it makes sense to move on to the next level of analysis by introducing the notion of direction. Now it is possible to identify strongly connected components (SCCs[2]). In terms of ownership networks, these patterns correspond to sets of corporations where every firm is connected to every other firm via a path of indirect ownership. Furthermore, these components may form bow-tie structures, akin to the topology of the World-Wide Web (Broder et al. 2000).[3] Figure 3.1 illustrates an idealized bow-tie topology. This structure reflects the flow of control, as every shareholder in the IN section exerts control and all corporations in the OUT section are controlled.

We find that roughly two thirds of the countries' ownership networks contain bow-tie structures (see also Sect. 4.2.6). Indeed, already at this level of analysis, previously observed patterns can be rediscovered. As an example, the countries with the

[2] A list of acronyms be found in front matter of this book.

[3] Recall Fig. 1.2 on p. 4 covering the same topic.

highest occurrence of (small) bow-tie structures are KR and TW, and to a lesser degree JP. A possible determinant is the well known existence of so-called business groups in these countries (e.g., the *keiretsu* in JP, and the *chaebol* in KR) forming a tightly-knit web of cross-shareholdings (see Sect. 1.2.2 and the references in Granovetter (1995); Feenstra et al. (1999)). For AU, CA, GB and US we observe very few bow-tie structures of which the largest ones however contain hundreds to thousands of corporations. It is an open question if the emergence of these mega-structures in the Anglo-Saxon countries is due to their unique "type" of capitalism (the so-called Atlantic or stock market capitalism, see Sect. 1.2.2 and the references in Dore (2002)), and whether this finding contradicts the assumption that these markets are characterized by the absence of business groups (Granovetter 1995).

Continuing with this line of research would lead to the question of how control is fragmented (e.g., investigations of the distribution of cluster sizes, cluster densities, etc.). Further analyzing this issue at the third level would require the weight of links and non-topological variables of the nodes to be considered as well. As our current interest is devoted to the question of how control is distributed, we do not further investigate the nature of the connected components. We ask instead what structures can be identified that reflect the concentration of control. Our proposed methodology answers this question by extracting the core structures of the ownership networks—the backbones—unveiling the seat of power in national stock markets (see Sect. 3.4).

### 3.3.2 Level 2: Extending the Notions of Degree

Basic measures in graph theory are the degree and strength, defined in Appendix B.2. However, the interpretation of these quantities is not always straightforward for real-world networks. In the case of ownership networks, as mentioned at the beginning of this section, there is no useful meaning associated with these values. In order to provide a more refined and appropriate description of weighted ownership networks, we introduce two quantities that extend the notions of degree and strength in a sensible way.

The first quantity to be considered reflects the relative importance of the neighbors of a vertex. More specifically, given a vertex $j$ and its incoming edges, we focus on the originating vertices of such edges. The idea is to define a quantity that captures the relative importance of incoming edges.

A way of measuring the number of prominent incoming edges is the concentration index defined in Eq. (2.122):

$$s_j := \frac{\left(\sum_{i=1}^{k_j^{in}} W_{ij}\right)^2}{\sum_{i=1}^{k_j^{in}} W_{ij}^2}. \tag{3.1}$$

See also Fig. 2.7 on p. 57. In the context of ownership networks, $s_j$ is interpreted as the effective number of shareholders of the stock $j$. Thus it can be understood as a measure of control from the point of view of a stock.

The second quantity introduced in Sect. 2.8.1, the fraction of control, measures the number of important outgoing edges of the vertices. It is defined in Eq. (2.123):

$$H_{ij} := \frac{W_{ij}^2}{\sum_{l=1}^{k_j^{in}} W_{lj}^2}. \tag{3.2}$$

Recall Fig. 2.8 on p. 59. For ownership networks, $H_{ij}$ represents the fraction of control shareholder $i$ has in the company $j$. A high value of $H_{ij}$, i.e., $H_{ij} \approx 1$, reflects the fact that $i$ is by far the most important destination vertex for the vertex $j$.

In a next step, the control index was given in Eq. (2.124):

$$h_i := \sum_{j=1}^{k_i^{out}} H_{ij}. \tag{3.3}$$

As shown in Fig. 2.8, this quantity is a way of measuring how important the outgoing edges of a node $i$ are with respect to its neighbors' neighbors. Within the ownership network setting, $h_i$ is interpreted as the effective number of stocks controlled by shareholder $i$.

The measures $s$ and $h$ introduced here are primarily used in the algorithm that extracts the backbone (see Sect. 3.4). Moreover, they are instrumental for the classification of the various national backbones, as explained in Sect. 3.4.4. Notably, $H_{ij}$ has its own *raison d'être* in an economics context. For an interpretation from this point of view, consult Sect. 2.8.2. In a nutshell, $H_{ij}$ is a new model to estimate control based on ownership relations. It extends the existing linear and threshold-based methods (see Sect. 2.7) by incorporating insights from game theory aimed at describing shareholder coalitions (consult Sect. 2.8).

### 3.3.3 Distributions of s and h

The measures $s$ and $h$ themselves can also already provide insights into the patterns of how ownership and control are distributed at a local level.

Figure 3.2 shows the probability density function (PDF) of $s_j$ for a selection of nine countries (the full sample is available on-line at: http://www.sg.ethz.ch/research/economic_networks/ownership_networks/online). There is a diversity in the shapes and ranges of the distributions to be seen. For instance, the distribution of GB reveals that many companies have more than 20 leading shareholders, whereas in IT few companies are held by more than five significant shareholders. Such country-specific

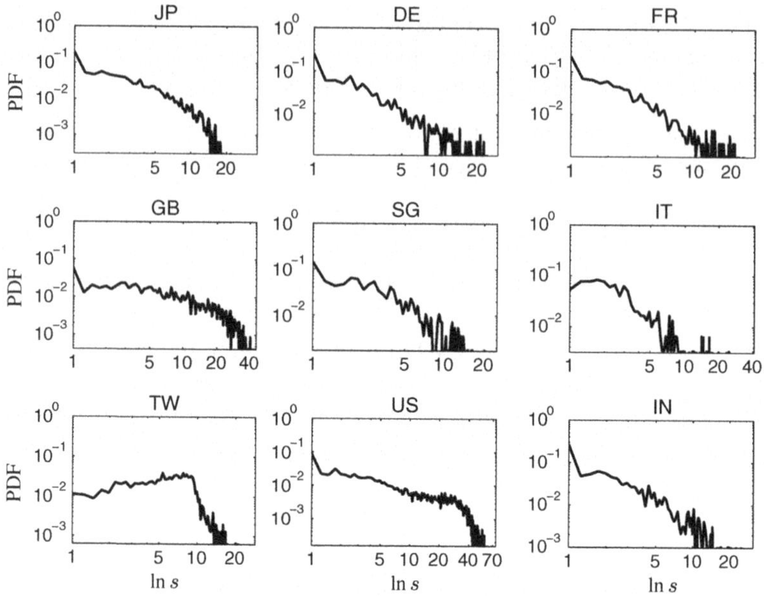


**Fig. 3.2** Probability distributions of $s_j$ for selected countries; PDF in log-log scale

signatures were expected to appear due to the differences in legal and institutional settings (e.g., law enforcement and protection of minority shareholders (La Porta et al. 1999)).

On the other hand, looking at the cumulative distribution function (CDF) of $k_i^{out}$ (shown for three selected countries[4] in the top panel of Fig. 3.3) a more uniform shape is revealed. The distributions range across two to three orders of magnitude. Hence some shareholders can hold up to a couple of thousand stocks, whereas the majority have ownership in less than 10. Considering the CDF of $h_i$, seen in the middle panel of Fig. 3.3, one can observe that the curves of $h_i$ display two regimes. This is true for nearly all analyzed countries, with a slight country-dependent variability. Notable exceptions are FI, IS, LU, PT, TN, TW, and VG. In order to understand this behavior it is useful to look at the PDF of $h_i$, shown in the bottom panel of Fig. 3.3. This uncovers a new systematic feature: the peak at the value of $h_i = 1$ indicates that there are many shareholders in the markets who's only intention is to control one single stock. This observation, however, could also be due to a database artefact as incompleteness of the data may result in many stocks having only one reported shareholder. In order to check that this result is indeed a feature of the markets, we constrain these ownership relations to the ones being bigger than 50 %, reflecting incontestable control. In a subsequent analysis we still observe this pattern in many countries (BM, CA, CH, DE, FR, GB, ID, IN, KY, MY, TH, US, ZA, and ES being the

[4] Again, the full sample is available on-line at: http://www.sg.ethz.ch/research/economic_networks/ownership_networks/online.

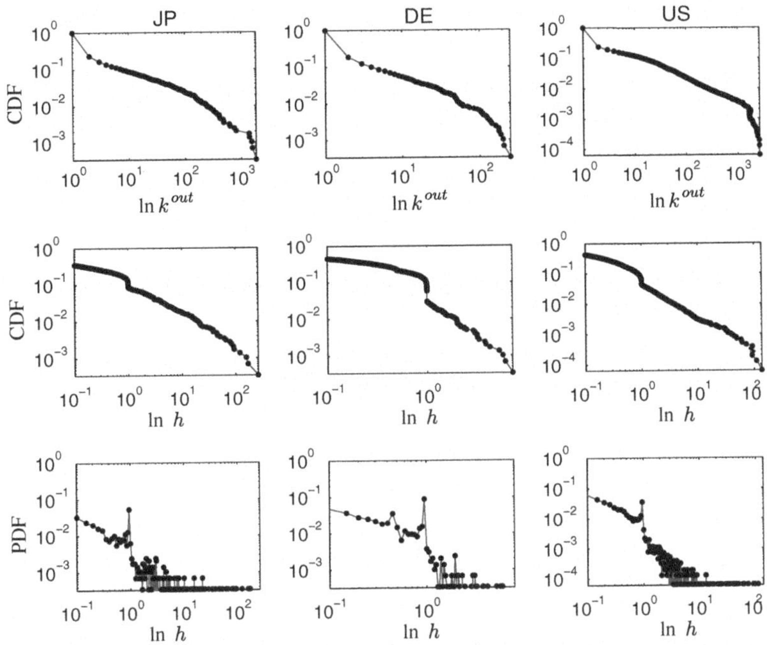


**Fig. 3.3** Various probability distributions for selected countries: (*top panel*) CDF plot of $k_i^{out}$; (*middle panel*) CDF plot of $h_i$; (*bottom panel*) PDF plot of $h_i$; all plots are in log-log scale

most pronounced). In addition, we find many such shareholders to be non-firms, i.e., people, families or legal entities, hardening the evidence for this type of exclusive control. This result emphasizes the utility of the newly defined measures to uncover relevant structures in the real-world ownership networks.

### 3.3.4 Level 3: Adding Non-Topological Values

The quantities defined in Eqs. (3.1) and (3.3) rely on the direction and weight of the links. However, they do not consider any non-topological state variables assigned to the nodes themselves. In our case of ownership networks, a natural choice is to use the market capitalization value of firms in thousand USD, $v_j$, as a proxy for their sizes. Hence $v_j$ will be utilized as the state variable in the subsequent analysis. In a first step, we address the question of how much wealth the shareholders own, i.e, the value in their portfolios.

As the percentage of ownership given by $W_{ij}$ is a measure of the fraction of outstanding shares $i$ holds in $j$, and the market capitalization of $j$ is defined by the number of outstanding shares times the market price, the following quantity reflects $i$'s portfolio value, also defined in Eq. (1.2):

$$p = Wv. \tag{3.4}$$

Extending this measure to incorporate the notions of control, we replace $W_{ij}$ in the previous equation with the fraction of control $H_{ij}$, defined in Eq. (3.2),[5] yielding the control value introduced in Eq. (2.119):

$$c = Hv. \tag{3.5}$$

A high $c_i$ value is indicative of the possibility to directly control a portfolio with a big market capitalization value.

In order to consider the effect of all direct and indirect paths in the network, the notion of integrated control was introduced in Eq. (2.126) of Sect. 2.9:

$$\tilde{\varsigma}^{\text{int}} = \tilde{H}v = \tilde{C}^{RM}v, \tag{3.6}$$

where, for a matrix $A$

$$\tilde{A} := (I - A)A^{-1}, \tag{3.7}$$

as seen in Eq. (2.42), or alternatively, Eq. (2.15). Recall that the conditions for Eq. (3.6) to have a solution are discussed in the paragraph containing Eq. (2.20) on p. 28 and the one following it.

This newly introduced quantity, measuring the economic value a shareholder can control in a network of ownership relations, is used in Sect. 3.4.1 to identify and rank the important shareholders.

As seen in Chap. 2, a lot of the methodology is devoted to modifying and correcting Eq. (3.6). In the scope of the current empirical analysis, it suffices to use the uncorrected integrated control. The reason being that only listed companies are analyzed. This means that the true ownership structures, also containing many non-listed companies, are only approximated. In this sense, the results given in this chapter should be understood as a first approach to the true network of corporations. The full analysis, building on a special algorithm explicitly constructing the network of all firms and shareholders around transnational corporations, is presented in Chap. 4. There, due to the complexity of the emerging ownership patterns, the corrections have to be considered.

Moreover, $\tilde{\varsigma}^{\text{int}}$ is only used as a proxy for the identification of important shareholders based on the potential control gained from the network. The method of computing cumulative control, seen in Sect. 3.4.1, then relies on the actual control value gained directly.

---

[5] Alternatively, see also Eq. (2.123).

## 3.4 Identifying the Backbone of Corporate Control

Based on the quantities introduced in the previous sections we are now in the position to proceed with the main aim of this chapter, which is to investigate the concentration of control in the ownership networks at a global level. This means, qualitatively, that we have to identify those shareholders who can be considered to be in control of the market. In detail, we develop an algorithm that extracts the core subnetwork from the ownership network, which we call the *backbone*. This structure consists of the smallest set of the most powerful shareholders that, collectively, are potentially able to control a predefined fraction of the market in terms of value.

To this aim, in Sect. 3.4.1, we introduce a ranking of the shareholders based on the value of the portfolio they control, as measured by the integrated control value $\tilde{\zeta}_i^{\text{int}}$. We are then able to compute how much value the top shareholders can potentially control, jointly, should they form a coalition. We call this notion *cumulative control*. Building on this knowledge, in Sect. 3.4.2, we extract the subnetwork of the most powerful shareholders and their (cumulatively) controlled stocks: the backbone. Section 3.4.3 presents a generalization of this backbone-extraction algorithm applicable to general weighted and oriented networks. The backbone structures of the analyzed countries are further investigated in Sect. 3.4.4. Different classification measures are introduced, allowing us to perform a cross-country analysis of how the control and value are globally distributed in the markets (Sect. 3.5.1) next to identifying who is holding the seat of power (Sect. 3.5.2).

### 3.4.1 Computing Cumulative Control

The first step of our methodology requires the construction of a Lorenz-like curve in order uncover the distribution of the value in a market. This concept was introduced in Sect. 2.10.

Here, on the $x$-axis we rank the shareholders according to their importance and report the fraction they represent with respect to the whole set of shareholder. The $y$-axis shows the corresponding percentage of controlled market value. In detail, we relate the fraction of shareholders ranked by their integrated control value $\tilde{\zeta}_i^{\text{int}}$, cf. Eqs. (3.2), and (3.6),[6] to the fraction of the total market value they collectively or cumulatively control.

In order to motivate the notion of cumulative control, some preliminary remarks are required. Using the integrated control value to rank the shareholders means that we implicitly assume control based on the integrated fraction of control $\widetilde{H}_{ij}$. This however is a potential value reflecting *possible* control. In order to identify the backbone, we take a very conservative approach to the question of what the *actual* control of a shareholder is. To this aim, we introduce a stringent threshold of 50 %. Any

[6] Alternatively, Eq. (2.126).

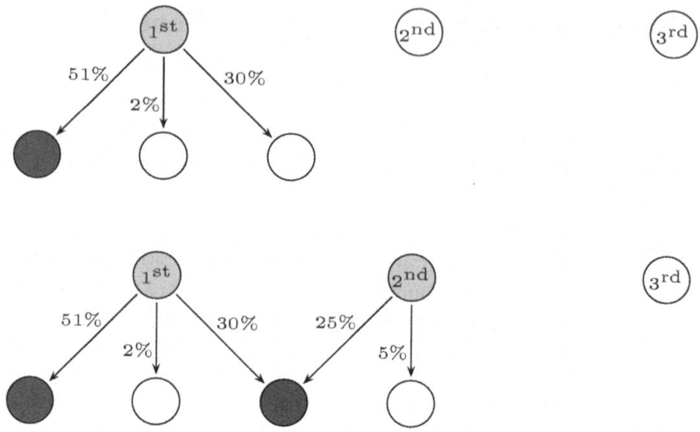


**Fig. 3.4** First steps in computing cumulative control: (*top panel*) selecting the most important shareholder (*light shading*) ranked according to the $\tilde{\varsigma}_i^{\text{int}}$-values and the portfolio of stocks owned at more than 50 % (*dark shading*); in the second step (*bottom panel*), the next most important shareholder is added; although there are now no new stocks which are owned directly at more than 50 %, cumulatively the two shareholder own an additional stock at 55 %

shareholder with an ownership percentage $W_{ij} > 0.5$ controls by default. This strict notion of control for a single shareholder is then generalized to apply to the cumulative control a group of shareholders can exert. Namely by requiring the sum of ownership percentages multiple shareholders have in a common stock to exceed the threshold of cumulative control. Its value is equivalently chosen to be 50 %.

We start the computation of cumulative control by identifying the shareholder having the highest $\tilde{\varsigma}_i^{\text{int}}$-value. From the portfolio of this holder, we extract the stocks that are owned at more than the said 50 %. In the next step, the shareholder with the second highest $\tilde{\varsigma}_i^{\text{int}}$-value is selected. Next to the stocks individually held at more than 50 % by this shareholder, additional stocks are considered, which are cumulatively owned by the top two shareholders at more than the said threshold value. See Fig. 3.4 for an illustrated example.

$U_{in}(n)$ is defined to be the set of indices of the stocks that are individually held above the threshold value by the $n$ selected top shareholders. Equivalently, $U_{cu}(n)$ represents the set of indices of the cumulatively controlled companies. It holds that $U_{in}(n) \cap U_{cu}(n) = \varnothing$. At each step $n$, the total value of this newly constructed portfolio, $U_{in}(n) \cup U_{cu}(n)$, is computed:

$$v_{cu}(n) := \sum_{j \in U_{in}(n)} v_j + \sum_{j \in U_{cu}(n)} v_j. \tag{3.8}$$

Equation (3.8) is in contrast to Eq. (3.4), where the total value of the stocks $j$ is multiplied by the ownership percentage $W_{ij}$. The computation of cumulative control

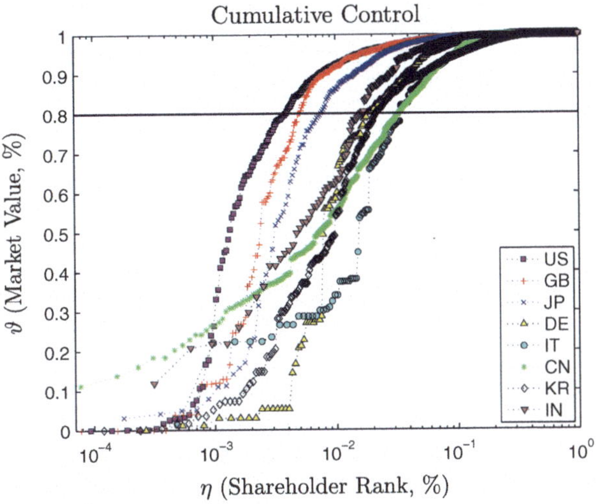


**Fig. 3.5** Concentration of control seen in different countries: fraction of shareholders $\eta$, sorted by descending (integrated) control value $\tilde{\xi}_i^{\text{int}}$, cumulatively controlling $\vartheta$ percent of the total market value; the *horizontal line* denotes a market value of 80 %; the diagram is in semi-log scale

is described in steps 1–7 (ignoring the termination condition in step 8) of Algorithm 1 on p. 80. Consult the next section for more details.

Let $n_{tot}$ be the total number of shareholders in a market and $v_{tot}$ the total market value. We normalize with these values, defining:

$$\eta(n) := \frac{n}{n_{tot}}, \qquad \vartheta(n) := \frac{v_{cu}(n)}{v_{tot}}, \tag{3.9}$$

where $\eta, \vartheta \in (0, 1]$.

In Fig. (3.5) these values are plotted against each other for a selection of countries, yielding the cumulative control diagram, akin to a Lorenz curve (with reversed $x$-axis). As an example, a coordinate pair with value $(10^{-3}, 0.2)$ reveals that the top 0.1 % of shareholders cumulatively control 20 % of the total market value. The top right corner of the diagram represents 100 % percent of the shareholders controlling 100 % of the market value, and the first data point in the lower left-hand corner denotes the most important shareholder of each country. Different countries show a varying degree of concentration of control.

Recall that for every shareholder the ranking is based on all paths of control of any length along the direction of the arrows (indirect control). For every such reachable stock the importance of its direct co-shareholders is considered (against the direction of the arrows). Therefore our analysis is based on a genuine network approach which allows us to gain crucial information on every shareholder, which would otherwise be

**Algorithm 1** $\mathcal{BB}(\tilde{\varsigma}_1^{\text{int}}, \ldots, \tilde{\varsigma}_n^{\text{int}}, \delta, \hat{\vartheta})$

1: $\tilde{\varsigma}^{\text{int}} \leftarrow sort\_descending(\tilde{\varsigma}_1^{\text{int}}, \ldots, \tilde{\varsigma}_n^{\text{int}})$
2: **repeat**
3: $\quad \varsigma \leftarrow get\_largest(\tilde{\varsigma}^{\text{int}})$
4: $\quad I \leftarrow I \cup index(\varsigma)$
5: $\quad PF \leftarrow stocks\_controlled\_by(I)$ (individually and cumulatively at more than $\delta$)
6: $\quad PFV \leftarrow value\_of\_portfolio(PF)$
7: $\quad \tilde{\varsigma}^{\text{int}} \leftarrow \tilde{\varsigma}^{\text{int}} \setminus \{\varsigma\}$
8: **until** $PFV \geq \hat{\vartheta} \cdot total\_market\_value$
9: $prune\_network(I, PF)$

undetectable. In contrast, most other empirical studies start their analysis from a set of important stocks (e.g., ranked by market capitalization). The methods of accounting for indirect control (see Sect. 3.3.4) are, if at all, only employed to detect the so-called ultimate owners of the stocks. Recall the discussion in the second paragraph of the subsection called "Existing Work" in Sect. 1.2.3.

Finally, note that although the identity of the individual controlling shareholders is lost due to the introduction of cumulative control, the emphasis lies on the fact that the controlling shareholders are present in the set of the first $n$ holders.

### 3.4.2 Extracting the Backbone

Once the curve of the cumulative control is known for a market, one can set a threshold for the percentage of jointly controlled market value, $\hat{\vartheta}$. This results in the identification of the percentage $\hat{\eta}$ of shareholders that theoretically hold the power to control this value, if they were to coordinate their activities in corresponding voting blocks. As mentioned, the subnetwork of these power-holders and their portfolios is called the backbone. Here we choose the value $\hat{\vartheta} = 0.8$, revealing the power-holders able to control 80 % of the total market value.

Algorithm 1 gives the complete recipe for computing the backbone. As inputs, the algorithm requires all the $\tilde{\varsigma}_i^{\text{int}}$-values, the threshold defining the level of (cumulative) control $\delta$, and the threshold for the considered market value $\hat{\vartheta}$. As mentioned in the last section, steps 1–7 are required for the cumulative control computation and $\delta$ is set to 0.5. Step 8 specifies the interruption requirement given by the controlled portfolio value being bigger than $\hat{\vartheta}$ times the total market value.

Finally, in step 9, the subnetwork of power-holders and their portfolios is pruned to eliminate weak links and further enhance the important structures: for each stock $j$, only as many shareholders are kept as the rounded value of $s_j$ indicates, i.e., the (approximate) effective number of shareholders. E.g., if $j$ has 5 holders but $s_j$ is roughly three, only the three largest shareholders are considered for the backbone. In effect, the weakest links and any resulting isolated nodes are removed.

### 3.4.3 Generalizing the Method of Backbone Extraction

Notice that our method can be generalized to any directed and weighted network in which (1) a non-topological real value $v_j \geq 0$ can be assigned to the nodes (with the condition that $v_j > 0$ for at least all the leaf-nodes in the network) and (2) an edge from node $i$ to $j$ with weight $W_{ij}$ implies that some of the value of $j$ is transferred to $i$. This was detailed in Sect. 2.4, where the methodology is reinterpreted with the notion of flow in networks (Sect. 2.4.1).

Returning to the backbone setting, let $U_0$ and $E_0$ be, respectively, the set of vertices and edges yielding the network. We define a subset $U \subseteq U_0$ of vertices on which we want to focus on (in the analysis presented earlier $U = U_0$). Let $E \subseteq E_0$ then be the set of edges among the vertices in $U$ and introduce $\hat{\vartheta}$, a threshold for the fraction of aggregate flow through the nodes of the network. If the relative importance of neighboring nodes is crucial, $H_{ij}$ is computed from $W_{ij}$ by the virtue of Eq. (3.2). Note that $H_{ij}$ can be replaced by any function of the weights $W_{ij}$ that is suitable in the context of the network under examination. We now solve Eq. (3.7) to obtain the integrated value $\widetilde{H}_{ij}$. This yields the quantitative relation of the indirect connections amongst the nodes. To be precise, it should be noted that in some networks the weight of an indirect connection is not correctly captured by the product of the weights along the path between the two nodes. In such cases one has to modify Eq. (2.43), the corresponding equation leading to $\widetilde{W}_{ij}$, accordingly.

The next step in the backbone extraction procedure is to identify the fraction of flow that is transferred by a subset of nodes. A systematic way of doing this was presented in Sect. 3.4.1 where we constructed the curve, $(\eta, \vartheta)$. A general recipe for such a construction is the following. On the $x$-axis all the nodes are ranked by their $\phi_i$-value in descending order and the fraction they represent with respect to size of $U$ is captured. The $y$-axis then shows the corresponding percentage of flow the nodes transfer. As an example, the first $k$ (ranked) nodes represent the fraction $\eta(k) = k/|U|$ of all nodes that cumulatively transfer the amount $\vartheta(k) = (\sum_{i=1}^{k} \phi_i)/\phi_{tot}$ of the total flow. Furthermore, $\hat{\eta}$ corresponds to the percentage of top ranked nodes that pipe the predefined fraction $\hat{\vartheta}$ of all the mass flowing in the whole network. Note that the procedure described in Sect. 3.4.1 is somewhat different. There we considered the fraction of the total value given by the direct successors of the nodes with largest $\varsigma_i^{\text{int}}$. This makes sense due to the special nature of the ownership networks under investigation, where every non-firm shareholder (root-node) is directly linked to at least one corporation (leaf-node), and the corporations are connected amongst themselves.

Consider the union of the nodes identified by $\hat{\eta}$ and their direct and indirect successors, together with the links amongst them. This is a subnetwork $\mathcal{B} = (U^B, E^B)$, with $U^B \subset U$ and $E^B \subset E$ that comprises, by construction, the fraction $\hat{\vartheta}$ of the total flow. This is already a first possible definition of the backbone of $(U, E)$. A discussion of the potential application of this procedure to other domains, and a more detailed description of the generalized methodology (along with specific refinements pertaining to the context given by the networks) is left for future work. Viable candidates are the world trade web (Serrano and Boguñá 2003; Garlaschelli and Loffredo 2004; Reichardt and White 2007; Fagiolo et al. 2008), food-webs (Garlaschelli et al. 2003), transportation networks (Kühnert et al. 2006), and credit networks (Battiston et al. 2007)

It should also be noted that in Sect. 3.4.1 we have introduced an additional threshold $\delta$ for the weights of the links which is needed in the context of corporate control. In the general case it can be set to zero. Returning to the specific context given by the data analyzed in this paper, one can vary the requirements that determine the backbone. For instance, one could focus on a predefined subset of listed companies, say the ten largest ones in the energy sector, and impose that the cumulative control over that set of stocks is $\hat{\vartheta} = 60\,\%$.

### 3.4.4 Defining Classification Measures

According to economists, markets differ from one country to another in a variety of respects, e.g., (La Porta et al. 1998, 1999) mentioned in Sect. 1.2.2. They may however not look too different if one restricts the analysis to the distribution of local quantities, and in particular to the degree, as shown in Sect. 3.3.3. In contrast, at the level of the backbones, i.e., the structures where most of the value resides, they can look strikingly dissimilar, as seen for instance in the case of CN and JP, shown in Fig. 3.6. In order to attempt a classification of these diverse structures, we will make use of indicators built on the same quantities used to construct the backbone. Performing a cross-country analysis for these indicators gives new insights into the characteristics of the global markets.

In detail, the properties we are interested in and want to unveil are the concentration of control and value, next to the frequency of widely held companies. In the following, straightforward metrics reflecting these characteristics are defined. Let $n_{st}$ and $n_{sh}$ denote the number of stocks and shareholders in a backbone, respectively. As $s_j$ measures the effective number of shareholders of a company, the average value

$$\overline{s} = \frac{\sum_{j=1}^{n_{st}} s_j}{n_{st}}, \tag{3.10}$$

is a good proxy characterizing the local patterns of ownership: the higher $\overline{s}$, the more dispersed the ownership is in the backbone, or the more common is the appearance of widely held firms. Furthermore, due to the construction of $s_j$, the metric $\overline{s}$ equivalently measures the local concentration of control.

In a similar vein, the average value

$$\overline{h} = \frac{\sum_{i=1}^{n_{sh}} h_i}{n_{sh}} = \frac{n_{st}}{n_{sh}}, \tag{3.11}$$

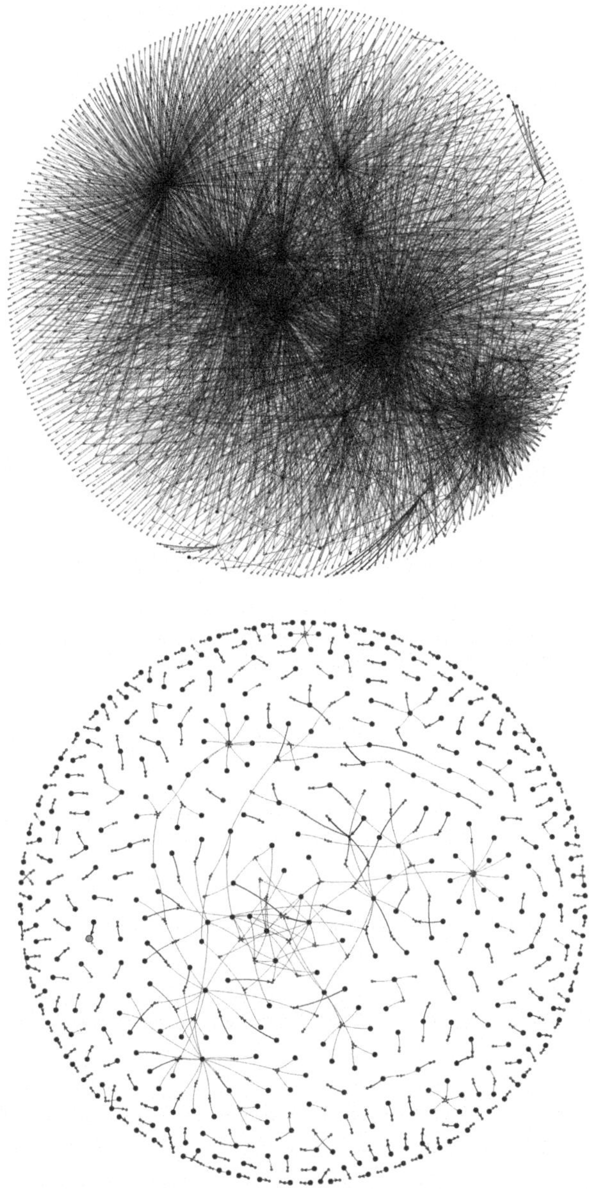

**Fig. 3.6** (*Top*) the backbone of JP; (*bottom*) the backbone of CN; the graph layouts are based on Geipel (2007); for the complete set of backbone layouts consult http://www.sg.ethz.ch/research/economic_networks/ownership_networks/online

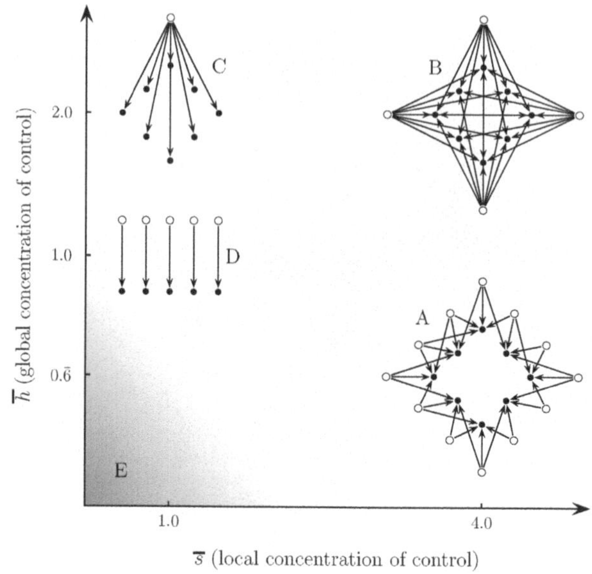


**Fig. 3.7** The map of control: illustration of idealized network topologies in terms of local dispersion of control ($x$-axis) versus global concentration of control ($y$-axis); shareholders and stocks are shown as *empty* and *filled* bullets, respectively; *arrows* represent ownership; consult the discussion in the text; see Fig. 3.9 for the empirical results

reflects the global distribution of control. A high value of $\overline{h}$ means that the considered backbone has very few shareholders compared to stocks, exposing a high degree of global concentration of control. It is worth noting that the values $n_{st}$ and $n_{sh}$ are derived from the backbone and are hence network-related measures.

Figure 3.7 anticipates the possible generic backbone configurations resulting from local and global distributions of control. Moving to the right-hand side of the $x$-axis the stocks have many shareholders (local dispersion of control), whereas stocks on the very left side have only one shareholder each. The $y$-axis depicts the global concentration of control, i.e., how many shareholders are controlling all the stocks in the market. Moving up the $y$-axis, the stocks are held by fewer and fewer shareholders. There is a consistency constraint on the coordinates that are allowed and region (E) is excluded. Possible network configurations are (A) many owners sharing many stocks, (B) few shareholders holding many stocks, (C) a single shareholder controlling all the stocks and (D) a situation with an equal number of shareholders, ownership ties and stocks. Note that (A) does not necessarily need to be a connected structure as many fragmented network configurations can result in such coordinates.

Recall that for the backbones to be constructed, a threshold for the controlled market value needed to be specified: $\hat{\vartheta} = 0.8$. In the cumulative control diagram seen in Fig. (3.5), this allows the identification of the number of shareholders being able to control this value. The value $\hat{\eta}$ reflects the percentage of power-holders corresponding

**Table 3.1** Classification measure values for a selection of countries; in Figs. 3.9 and 3.10 these values are plotted for all analyzed countries

| | $\eta'(\%)$ | $\overline{s}$ | $\overline{h}$ |
|---|---|---|---|
| AU | 0.82 | 5.45 | 2.79 |
| CA | 3.32 | 3.04 | 4.97 |
| CH | 5.97 | 2.91 | 0.66 |
| CN | 9.21 | 1.32 | 0.90 |
| DE | 3.22 | 2.76 | 0.82 |
| FR | 3.96 | 2.65 | 0.83 |
| GB | 0.89 | 8.60 | 5.05 |
| IN | 5.27 | 2.15 | 3.92 |
| IT | 6.10 | 1.62 | 0.82 |
| JP | 1.93 | 2.48 | 34.26 |
| KR | 2.25 | 2.39 | 0.94 |
| TW | 5.00 | 2.98 | 0.58 |
| US | 0.56 | 8.56 | 15.39 |

to $\hat{\vartheta}$. To adjust for the variability introduced by the different numbers of shareholders present in the various national stock markets, we chose to normalize $\hat{\eta}$. Let $n_{100}$ denote the smallest number of shareholders controlling 100 % of the total market value $v_{tot}$, then

$$\eta' := \frac{\hat{\eta}}{n_{100}}. \tag{3.12}$$

A small value for $\eta'$ means that there will be very few shareholders in the backbone compared to the number of shareholders present in the whole market, reflecting that the market value is extremely concentrated in the hands of a few shareholders. In essence, the metric $\eta'$ is an emergent property of the backbone extraction algorithm and mirrors the global distribution of the value.

To summarize:

- $\overline{s}$ reflects local dispersion of control (at first-neighbor level, insensitive to value);
- $\overline{h}$ is an indicator of the global concentration of control (an integrated measure, i.e., derived by virtue of Eq. (3.7), at second-neighbor level, insensitive to value);
- $\eta'$ is a global measure of the concentration of market value (an emergent quantity).

Table 3.1 shows the empirical values of these quantities for a selection of countries. In the following, the results of a cross-country analysis for the classification measures are given.

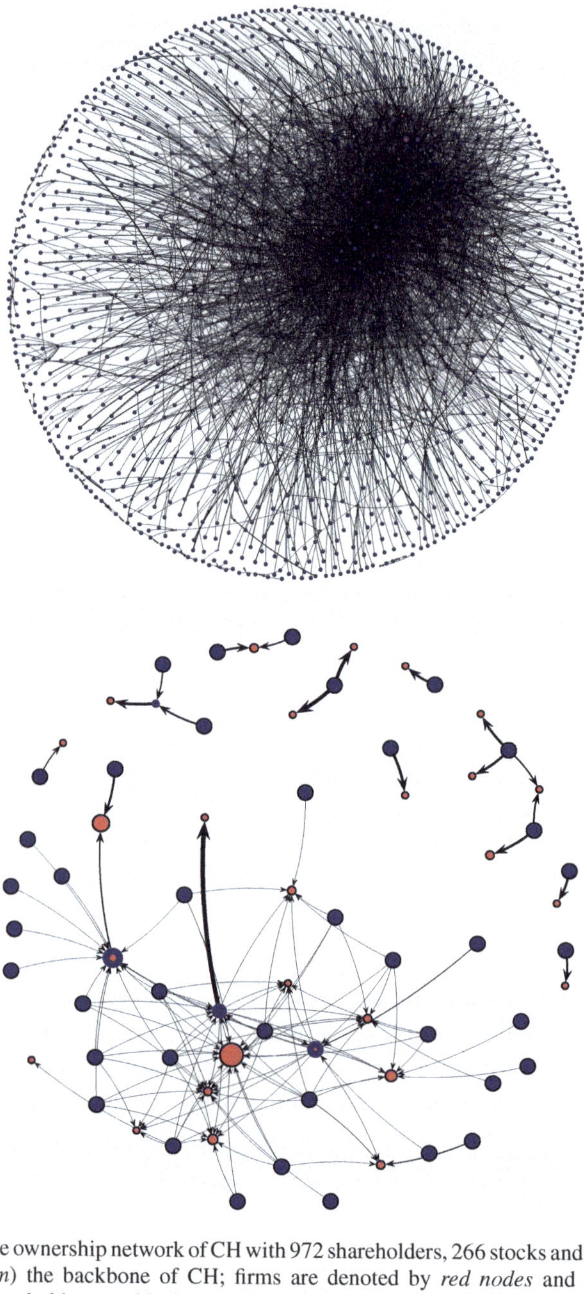

**Fig. 3.8** (*Top*) the ownership network of CH with 972 shareholders, 266 stocks and 4,671 ownership relations; (*bottom*) the backbone of CH; firms are denoted by *red nodes* and sized by market capitalization, shareholders are *black*, whereas firms owning stocks themselves are represented by *red nodes* with *thick* bounding *circles*, *arrows* are weighted by the percentage of ownership value; the graph layouts are based on Geipel (2007)

## 3.5 Analyzing the Backbones

In the last section, an algorithm for extracting the backbones of national markets, and measures reflecting their key characteristics, were given. But how relevant are these methods and how much of the properties of the real-world ownership networks they describe are captured?

Figure 3.8 shows the layout for the CH ownership network and the backbone, respectively. There is a big reduction in complexity by going to the backbone. Looking at the stocks left in the backbone, it is indeed the case that the important corporations reappear (recall that the algorithm selected the shareholders). We find a cluster of shareholdings linking, for instance, Nestlé, Novartis, Roche Holding, UBS, Credit Suisse Group, ABB, Swiss Re, Swatch, and JPMorgan Chase & Co. features as most important controlling shareholder. The descendants of the founding families of Roche (Hoffmann and Oeri) are the highest ranked Swiss shareholders at position four. UBS follows as dominant Swiss shareholder at rank seven.

The backbone extraction algorithm is also a good test for the robustness of market patterns. The bow-tie structures (discussed in Sect. 3.3.1) in JP, KR, TW vanish or are negligibly small in their backbones, whereas in the backbones of the Anglo-Saxon countries (and as an outlier SE) one sizable bow-tie structure survives. This emphasizes the strength and hence the importance of these patterns in the markets of AU, CA, GB and the US.

But what about some of the findings in ownership patterns that have been previously reported in the literature? To see if we can recover some known observations, we analyze the empirical values for the "Widely Held" index defined in La Porta et al. (1999), where a value of one is assigned if there are no controlling shareholders, and zero if all firms in the sample are controlled. There is a threshold introduced, beyond which control is said to occur: the study is done with a 10 and 20 % cutoff value. We find a 76.6 % correlation (and a $p$-value for testing the hypothesis of no correlation of $3.2 \cdot 10^{-6}$) between $\overline{s}$ in the backbone and the 10 % cutoff "Widely Held" index for the 27 countries it is reported for. The correlation of $\overline{s}$ in the countries' whole ownership networks is 60.0 % ($9.3 \cdot 10^{-4}$). For the 20 % cutoff, the correlation values are smaller. These relations should however be handled with care, as the study (La Porta et al. 1999) is restricted to the 20 largest firms (in terms of market capitalization) in the analyzed countries and there is a twelve-year lag between the datasets in the two studies. Nevertheless, it is a reassuring sign to find such a high correlation with older proxies for the occurrence of widely held firms.

Having established that the backbones indeed successfully comprise important structures of the markets, and showing that one of the classification methods we propose confirms known results, we can proceed to investigate novel aspects of the ownership networks. As frequently mentioned in this paper, the lack of existing network-oriented analysis of the financial architecture of corporations in national markets leaves one question unaddressed: what is the *global* concentration of control?

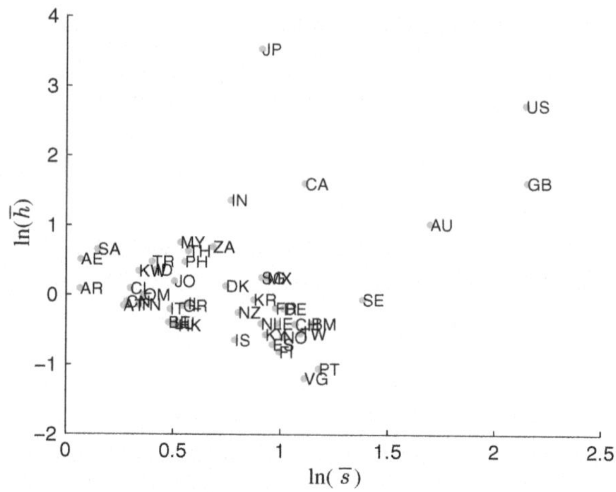


**Fig. 3.9** Map of control: local dispersion of control, $\overline{s}$, is plotted against global concentration of control, $\overline{h}$, for 48 countries

## *3.5.1 Global Concentration of Control*

We utilize the measures defined in Eqs. (3.10)–(3.12), to classify the 48 backbones. To recapitulate, $\overline{s}$ is a local measure for the dispersion of control. A large value indicates a high presence of widely held firms. $\overline{h}$ is a second-neighbor quantity sensitive to the concentration of global control. Large values are indicative that the control of many stocks resides in the hands of very few shareholders. Finally, $\eta'$ is a global variable related to the (normalized) percentage of shareholders in the backbone. It hence measures the concentration of value in a market, as a low number means that very few shareholders are able to control 80 % of the market value.

In Fig. 3.9 the log-values of $\overline{s}$ and $\overline{h}$ are plotted against each other. The $\overline{s}$-coordinates of the countries are as expected (La Porta et al. 1999): to the right we see the presence of widely held firms (i.e., the local dispersion of control) for the Anglo-Saxon countries AU, GB and the US. FR, IT, JP are located to the left, reflecting more concentrated local control. However, what is astonishing is that there is a counterintuitive trend to be observed in the data: the more local control is dispersed, the higher the global concentration of control becomes. In essence, what looks like a democratic distribution of control from close up, by taking a step back, actually turns out to warp into highly concentrated control in the hands of very few shareholders. On the other hand, the local concentration of control is in fact widely distributed amongst many controlling shareholder. Comparing with Fig. 3.7, where idealized network configurations are illustrated, we conclude that the empirical patterns of local and global control range from the type (B) to type (D), with JP combining local and global concentration of control. Interestingly, type (A) and (C) constellations are not observed in the data.

In Fig. 3.10 the log-values of $\overline{s}$ and $\eta'$ are depicted. What we concluded in the last paragraph for control is also true for the market value: the more the control is locally

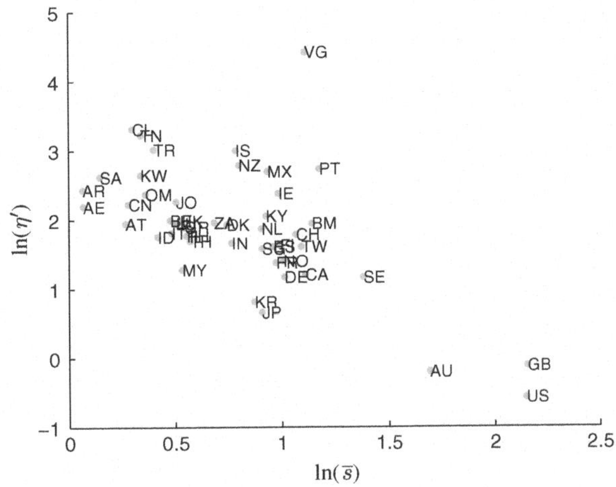


**Fig. 3.10** Map of market value: local dispersion of control, $\overline{s}$, is plotted against global concentration of market value, $\eta'$, for 48 countries

dispersed, the higher the concentration of value that lies in the hands of very few controlling shareholders, and vice versa.

We can also compare the $\overline{s}$ and $\overline{h}$ values measured for the backbones with the corresponding values of the total ownership networks, $\overline{s}_{tot}$ and $\overline{h}_{tot}$. We find that

$$\overline{s} < \overline{s}_{tot}. \tag{3.13}$$

This fact, that the widely held firms are less often present in the national backbones, means that the important shareholders (able to control 80 % of the market value) only infrequently invest in corporations with dispersed ownership. Note that the pruning scheme used in the construction of the backbone (introduced at the end of Sect. 3.4.2) approximates $s_j$ to the nearest integer. This can reduce the value of $\overline{s}$ in the backbone maximally by 0.5. In contrast, in our data (with the exception of ES) the relation $\overline{s}_{tot} - \overline{s} \gg 0.5$ holds, indicating that there is indeed a tendency of power-holders to avoid widely held firms, accounting for their less frequent appearance in the backbones.

We also find that

$$\overline{h} > \overline{h}_{tot}. \tag{3.14}$$

This means that there is a higher level of global control in the backbone, again implying that widely held firms occur less often in the backbone. In addition, looking at the ranges of $\overline{h}_{tot} \in [0.06, 1.09]$ and $\overline{h} \in [0.3, 34.26]$, reveals a higher cross-country variability in the backbone. In essence, the algorithm for extracting the backbone in fact amplifies subtle effects and unveils key structures.

We realize that the two figures discussed in this section open many questions. Why are there outliers to be observed: JP in Fig. 3.9 and VG in Fig. 3.10? What does it mean to group countries according to their $\overline{s}$, $\overline{h}$ and $\eta'$ coordinates and what does proximity imply? What are the implications for the individual countries? We hope to address such and similar questions in future work.

### 3.5.2 *The Seat of Power*

Having identified the important shareholders in the global markets, it is now also possible to address the following questions. Who holds the power in an increasingly globalized world? How important are individual people compared to the sphere of influence of multinational corporations? How eminent is the influence of the financial sector? By look in detail at the identity of the power-holders featured in the backbones, we address these issues next.

If one focusses on how often the same power-holders appear in the backbones of the 48 countries analyzed, it is possible to identify the global power-holders. Following is a top-ten list, comprised of the companies name, activity, country the headquarter is based in, and ranked according to the number of times it is present in different countries' backbones:

1. The Capital Group Companies; investment management; US; 36;
2. Fidelity Management & Research; investment products and services; US; 32;
3. Barclays PLC; financial services provider; GB; 26;
4. Franklin Resources; investment management; US; 25;
5. AXA; insurance company; FR; 22;
6. JPMorgan Chase & Co.; financial services provider; US; 19;
7. Dimensional Fund Advisors; investment management; US; 15;
8. Merrill Lynch & Co.; investment management; US; 14;
9. Wellington Management Company; investment management; US; 14;
10. UBS; financial services provider; CH; 12.

Note that this list represent a first attempt in identifying the global power-holders. In in Table E.1 of Appendix E we present a refined list based on the computation of corporate control in the true global network, presented in Chap. 4.

Next to the dominance of US American companies we find: Barclays PLC (GB), AXA (FR) and UBS (CH), Deutsche Bank (DE), Brandes Investment Partners (CA), Société Générale (FR), Credit Suisse Group (CH), Schroders PLC (GB), Allianz (DE) in the top 21 positions. The government of Singapore is at rank 25. HSBC Holdings PLC (HK/GB), the world's largest banking group, only appears at position 26. In addition, large multinational corporations outside of the finance and insurance industry do not act as prominent shareholders and only appear in their own national countries' backbones as controlled stocks. For instance, Exxon Mobil, Daimler Chrysler, Ford Motor Company, Siemens, Unilever.

Individual people do not appear as multinational power-holders very often. In the US backbone, we find one person ranked at ninth position: Warren E. Buffet. William Henry Gates III is next, at rank 26. In DE the family Porsche/Piech and in FR the family Bettencourt are power-holders in the top ten. For the tax-haven KY one finds Kao H. Min (who is placed at number 140 in the Forbes 400 list) in the top ranks.

The prevalence of multinational financial corporations in the list above is perhaps not very surprising. For instance, Capital Group Companies is one of the world's largest investment management organizations with assets under management in excess of one trillion USD. However, it is an interesting and novel observation that all the above mentioned corporations appear as prominent *controlling* shareholders simultaneously in many countries. We are aware that financial institutions such as mutual funds may not always seek to exert overt control. This is argued, for instance, for some of the largest US mutual funds when operating in the US (Davis 2008), on the basis of their propensity to vote against the management (although, the same mutual funds are described as exerting their power when operating in Europe). However, to our knowledge, there are no systematic studies about the control of financial institutions over their owned companies world-wide. To conclude, one can interpret our quantitative measure of control as potential power (namely, the probability of achieving one's own interest against the opposition of other actors (Weber 1997)). Given these premises, we cannot exclude that the top shareholders having vast potential power do not globally exert it in some way.

## 3.6 Summary and Conclusion

We have developed a methodology to identify and extract the backbone of complex networks that are comprised of weighted and directed links, and nodes to which a scalar quantity is associated. We interpret such networks as systems in which mass is created at some nodes and transferred to the nodes upstream. The amount of mass flowing along a link from node $i$ to node $j$ is given by the scalar quantity associated with the node $j$, times the weight of the link, $W_{ij}\, v_j$. The backbone corresponds to the subnetwork in which a preassigned fraction of the total flow of the system is transferred.

Applied to ownership networks, the procedure identifies the backbone as the subnetwork where most of the control and the economic value resides. In the analysis the nodes are associated with non-topological state variables given by the market capitalization value of the firms, and the indirect control along all ownership pathways is fully accounted for. We ranked the shareholders according to the value they can control and we constructed the subset of shareholders which collectively control a given fraction of the economic value in the market. In essence, our algorithm for extracting the backbone amplifies subtle effects and unveils key structures. We further introduced some measures aimed at classifying the backbones of the different markets in terms of local and global concentration of control and value. We find that each level of detail in the analysis uncovers new features in the ownership networks. Incorporating the direction of links in the study reveals bow-tie structures in the network. Including value allows us to identify who is holding the power in the global stock markets.

With respect to other studies in the economics literature, next to proposing a new model for estimating control from ownership, we are able to recover previously observed patterns in the data, namely the frequency of widely held firms in the various countries studied. Indeed, it has been known for over 75 years that the Anglo-Saxon countries have the highest occurrence of widely held firms (Berle and Means 1932). This statement, that the control of corporations is dispersed amongst many shareholders, invokes the intuition that there exists a multitude of owners that only hold a small amount of shares in a few companies. However, in contrast to such intuition, our main finding is that a local dispersion of control is associated with a global concentration of control and value. This means that only a small elite of shareholders controls a large fraction of the stock market, without ever having been previously systematically reported on. Some authors have suggested such a result by observing that a few big US mutual funds managing personal pension plans have become the biggest owners of corporate America since the 1990s (Davis 2008). On the other hand, in countries with local concentration of control (mostly observed in European states), the shareholders tend to only hold control over a single corporation, resulting in the dispersion of global control and value. Finally, we also observe that the US financial sector holds the seat of power at an international level. It will remain to be seen, if the continued unfolding of the current financial crisis will tip this balance of power, as the US financial landscape faces a fundamental transformation in its wake.

For an in-depth discussion of the relevance of our work and a summary of the overall implications, consult Sect. 3.6.

## References

P.W. Anderson, More is different. Science **177**(4047), 393–396 (1972)

A. Barrat, M. Barthelemy, R. Pastor-Satorras, A. Vespignani, The architecture of complex weighted networks. Proc. Natl. Acad. Sci. **101**, 3747 (2004)

S. Battiston, D. Delli Gatti, M. Gallegati, B. Greenwald, J.E. Stiglitz, Credit chains and bankruptcy propagation in production networks. J. Econ. Dyn. Control **31**(6), 2061–2084 (2007)

A. Berle, G. Means, *The Modern Corporation and Private Property* (Mac Millan, New York, 1932)

A. Broder, R. Kumar, F. Maghoul, P. Raghavan, S. Rajagopalan, S. Stata, A. Tomkins, J. Wiener, Graph structure in the web. Comput. Netw. **33**, 309 (2000)

G.F. Davis, A new finance capitalism? mutual funds and ownership re-concentration in the united states. Eur. Manag. Rev. **5**(1), 11–21 (2008)

R. Dore, Stock market capitalism vs. welfare capitalism. New Polit. Econ. **7**(1), 115–127 (2002)

G. Fagiolo, Clustering in complex directed networks. Phys. Rev. E **76**(2), 26107 (2007)

G. Fagiolo, J. Reyes, S. Schiavo, On the topological properties of the world trade web: a weighted network analysis. Physica A **387**, 3868–3873 (2008)

R.C. Feenstra, T.-H. Yang, G.G. Hamilton, Business groups and product variety in trade: evidence from South Korea. Taiwan and Japan. J. Int. Econ. **48**(1), 71–100 (1999)

D. Garlaschelli, M.I. Loffredo, Fitness-dependent topological properties of the world trade web. Phys. Rev. Lett. **93**(18), 188701 (2004)

D. Garlaschelli, G. Caldarelli, L. Pietronero, Universal scaling relations in food webs. Nature **423**(6936), 165–168 (2003)

D. Garlaschelli, S. Battiston, M. Castri, V.D.P. Servedio, G. Caldarelli, The scale-free topology of market investments. Phys. A **350**, 491–499 (2005)

M.M. Geipel, Self-organization applied to dynamic network layout. Int. J. Mod. Phys. C **18**(10), 1537–1549 (2007)

J.B. Glattfelder, S. Battiston, Backbone of complex networks of corporations: the flow of control. Phys. Rev. E **80**(3), 36104 (2009)

M. Granovetter, *Industrial and Corporate Change, Chapter Coase Revisited: Business Groups in the Modern Economy* (Oxford University Press, New York, 1995)

C. Kühnert, D. Helbing, G.B. West, Scaling laws in urban supply networks. Phys. A **363**, 96–103 (2006)

R. La Porta, F. Lopez de Silanes, A. Shleifer, Corporate ownership around the world. J. Finance **54**(2), 471–517 (1999)

R. La Porta, F. Lopez de Silanes, A. Shleifer, R.W. Vishny, Law and finance. J. Polit. Econ. **106**(6), 1113–1155 (1998)

J. Reichardt, D.R. White, Role models for complex networks. Eur. Phys. J. B **60**(2), 217–224 (2007)

M.Á. Serrano, M. Boguñá, Topology of the world trade web. Phys. Rev. E **68**(1), 15101 (2003)

M. Serrano, M. Boguñá, A. Vespignani, Extracting the multiscale backbone of complex weighted networks. Proc. Natl. Acad. Sci. **106**(16), 6483 (2009)

M. Weber, *The Theory of Social and Economic Organization* (Free Press, London, 1997)

# Chapter 4
# The Network of Global Corporate Control

*No one expects actually to deduce any principles of biology, psychology or politics from those of physics. The reason why higher-level subjects can be studied at all is that under special circumstances the stupendously complex behavior of vast numbers of particles resolves itself into a measure of simplicity and comprehensibility. This is called emergence.*

(D. Deutschin in Deutsch 1998, p. 20)

This chapter is based on the paper Vitali et al. (2011) and partly on Vitali (2010). The copyright of some of the images belongs to *PLoS ONE*. Note that in order to make the chapter self-consistent and self-supporting, some redundancies with Chaps. 1 and 2 are taken into account.

## 4.1 Introduction

In the last chapter we introduced an empirical network study based on the ownership network of listed companies in various countries (see Sect. 3.2). The analysis comprised 48 country networks, totalling 131018 nodes (24877 corporations or stocks and 106141 shareholding entities that cannot be owned themselves) and 545896 links.

In this chapter, we present an analysis of the global ownership network of shareholders, firms and subsidiaries located around the world's transnational corporations (TNCs).[1]

This vast network has to be algorithmically constructed. As it has never before been analyzed, we first describe some of its structure, properties and regularities before analyzing the nature of global control. In other words, Sect. 4.2 focusses on the Level 1 and 2 network analysis, while Sect. 4.3 gives the full 3-level analysis.

---

[1] A list of acronyms can be found in front matter of this book.

J. B. Glattfelder, *Decoding Complexity*, Springer Theses,
DOI: 10.1007/978-3-642-33424-5_4, © Springer-Verlag Berlin Heidelberg 2013

## 4.2 The Anatomy of the Global Corporate Ownership Network

### 4.2.1 The TNC Network Construction

The Orbis 2007 marketing database[2] comprises about 37 million economic actors, both physical persons and firms located in 194 countries, and roughly 13 million directed and weighted ownership links (equity relations). This dataset is intended to track control relationships rather than patrimonial relationships. Whenever available, the percentage of ownership refers to shares associated with voting rights.

The definition of TNCs given by OECD (2000) states that they

> *[...] comprise companies and other entities established in more than one country and so linked that they may coordinate their operations in various ways, while one or more of these entities may be able to exercise a significant influence over the activities of others, their degree of autonomy within the enterprise may vary widely from one multinational enterprise to another. Ownership may be private, state or mixed.*

Accordingly, we select those companies which hold at least 10 % of shares in companies located in more than one country. However, many subsidiaries of large TNCs themselves fulfill this definition of TNC (e.g., The Coca-Cola Company owns Coca-Cola Hellenic Bottling Company which in turn owns Coca-Cola Beverages Austria). Since for each multinational group we are interested in retaining only one representative, we exclude from the selection the companies for which the so-called ultimate owner (i.e., the owner with the highest share at each degree of ownership upstream of a company) is quoted in a stock market. In substitution, we add the quoted ultimate owner to the list (if not already included). In the example above, this procedure identifies only the Coca-Cola Company as a TNC. Overall we obtain a list of 43060 TNCs located in 116 different countries, with 5675 TNCs quoted in stock markets.

Starting from this list of TNCs, we explore recursively the neighborhood of companies in the whole database. First, we proceed downstream of the TNCs with a breadth-first search (BFS) and we identify all companies participated directly and indirectly by the TNCs. See Fig. 4.1. We then proceed in a similar way upstream identifying all direct and indirect shareholders of the TNCs. The resulting network can be divided into three classes of nodes: the TNCs, shareholders (SHs) and participated companies (PCs), as shown in Fig. 4.2. Observe that it may be possible to reach a PC from several TNCs, or to reach a TNC from several SHs. In other words, paths proceeding downstream or upstream of the TNCs may overlap, giving rise to connected components of various sizes.

The class of PCs contains direct and indirect subsidiaries of the TNCs, as well as other companies owned with smaller shares. On the other hand, the class of SHs contains both physical persons next to firms holding shares directly and indirectly. This procedure singles out, for the first time to our knowledge, the network of all the ownership pathways originating from and pointing to TNCs, as illustrated in Fig. 4.2. The

---

[2] Consult http://www.bvdep.com/orbis.html.

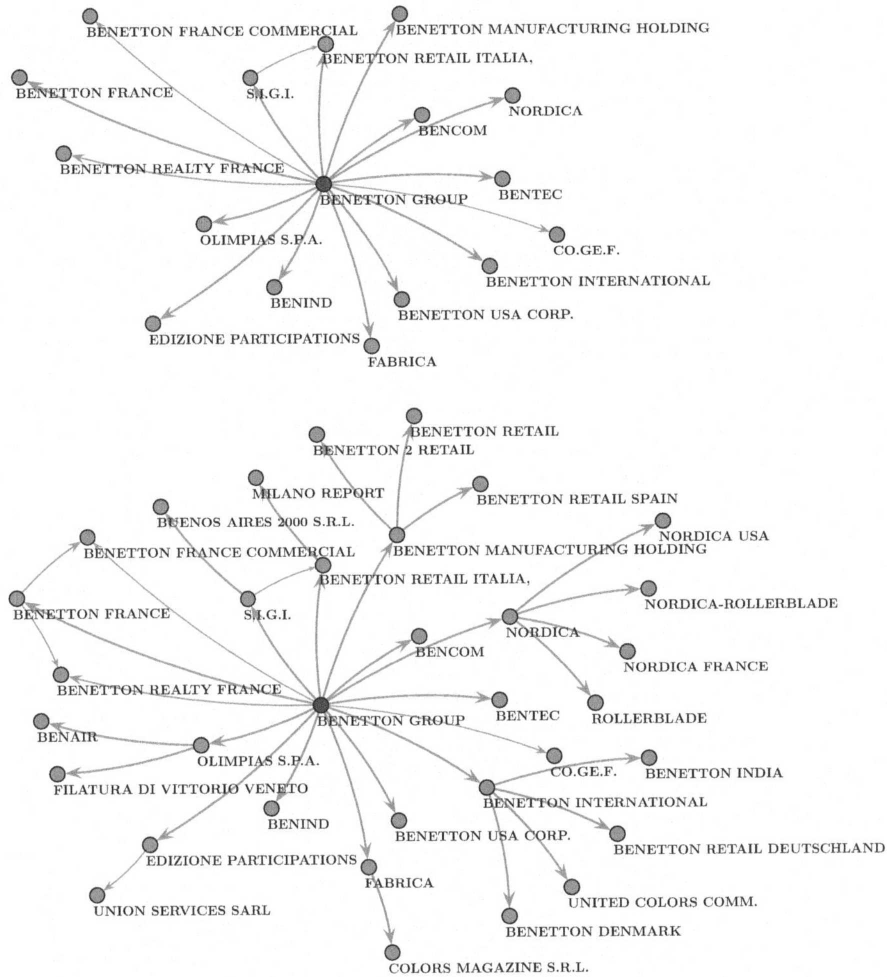


**Fig. 4.1** Illustration of the first two steps in the recursive exploration downstream of a TNC: starting from "Benetton Group" the BFS explores all the direct neighbors (*top panel*), and then the neighbors' neighbors (*bottom panel*)

dataset we extract in this way and analyze consists of 600508 economic entities, with 1006987 ownership relations among them, their geographical location, industrial sector and, for the TNCs only, operating revenue as a proxy of intrinsic company value.

### 4.2.2 Level 1 & 2 Network Measures: Degree and Strength

The study of the node degree (see Appendix B.2) refers to the distribution of the number of in-going and out-going relations. The number of outgoing links of a

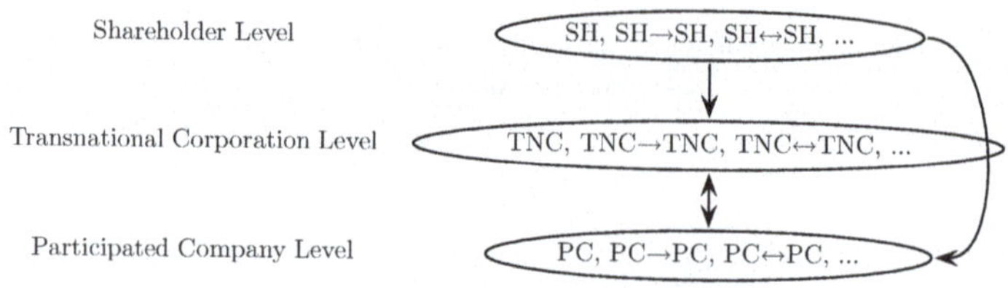


**Fig. 4.2** General structure of the TNC network with the three types of economic actors: 77456 SHs, 43060 TNCs and 479992 PCs; the network contains in total 600508 nodes, and 1006987 links; links are mainly from the TNCs to the PCs and amongst the PCs themselves

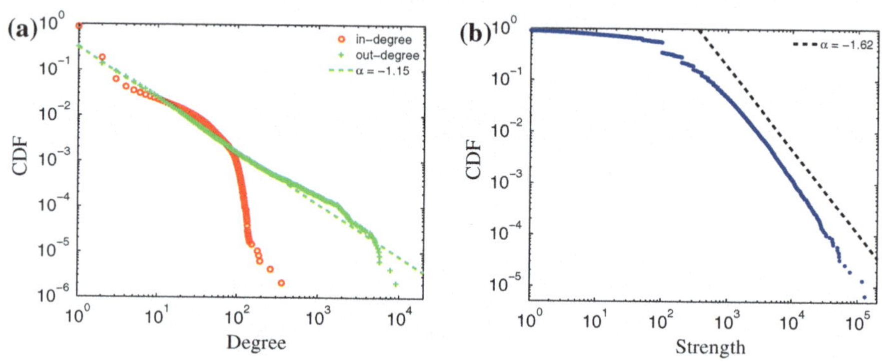


**Fig. 4.3** **a** Cumulative distribution function of the in- and out-degree of the nodes in the largest connected component (LCC) of the network (log-log scale); the power-law exponent for the corresponding probability density function of the out-degree is estimated to be $-2.15$ ($= \alpha - 1$); **b** cumulative distribution function of the node strength in the LCC (log-log scale); as a reference, a power-law with an exponent of $-1.62$ is displayed

node corresponds to the number of firms in which a shareholder owns shares. It is a rough measure of the portfolio diversification. The in-degree corresponds to the number of shareholders owning shares in a given firm. It can be thought of as a proxy for control fragmentation. In the TNC network, the out-degree can be approximated by a power law distribution with the exponent $-2.15$, see Fig. 4.3a. The majority of the economic actors points to few others resulting in a low out-degree. At the same time, there are a few nodes with a very high out-degree (the maximum number of companies owned by a single economic actor exceeds 5000 for some financial companies). On the other hand, the in-degree distribution, i.e., the number of shareholders of a company, behaves differently: the frequency of nodes with high in-degree decreases very fast. This is due to the fact that the database cannot provide all the shareholders of a company, especially those that hold only very small shares.

Next to the study of the node degree, we also investigate the strength which is defined as $\sum_j W_{ij}$, that is, the sum of all the weighed participations a company $i$ has in other companies $j$, see Fig. 4.3b. It is a measure of the weight connectivity and gives information on how strong the ownership relationships of each node are.

### 4.2.3 Unveiling the Topology at Level 1 & 2

In graph theory, (weakly) connected components (CCs) represent a clustering of nodes, where each node is connect to every other node through a chain of undirected links. In a directed network, a CC can be identified if one ignores the direction of the links.

In contrast, as introduced in Sect. 1.1, strongly connected components (SCCs) are sets of nodes that are all reachable from each other following chains of directed links. Every SCC can be understood as the core of a bow-tie structure, as seen in Figs. 1.2 and 3.1 on p. 4 and 71, respectively. It describes a core-periphery structure, with an incoming (IN) and outgoing (OUT) segment.

Ownership relations between companies create formal ties among them. In a SCC, as all firms are linked to each other via ownership pathways, they all own each other indirectly to some extent. See also Sect. 1.2.2. Although in a CC firms can reach each other only if one ignores the direction of the ownership links, this is still a situation of interest from an economics point of view. For instance, the flow of knowledge and information is not restricted by the direction of the link. Moreover, the number and the size distribution of the CCs provides a measure of the fragmentation of the market.

### 4.2.4 Weakly Connected Components

We find that the TNC network consists of 23825 CCs. A majority of the nodes (77 %) belong to the largest connected component (LCC) with 463006 economic actors and 889601 relations. The remaining nodes belong to CCs with sizes at least 2000 times smaller. The second largest CC contains 230 nodes and 90 % of the CC have less than 10 nodes. In Fig. 4.4 the distribution of the CC sizes is shown. So, although the whole TNC network is overall very fragmented, the top TNCs by economic value are all part of the LCC, containing slightly more than 3/4 of all the nodes.

Only 36 % of the TNCs are located in the LCC. However, they account for 94.2 % of the total TNC operating revenue. The remaining 1/4 of the nodes, among them the remaining 64 % of the TNCs with less than 6 % of the operating revenue, are distributed among 23824 other connected components (OCCs).

From a geographical point of view, the LCC includes companies from 191 countries. Of these nodes, 15491 are TNCs from 83 different countries. The firms that are PCs are much more numerous (399696) but are located in only 38 countries. Finally, there are 47819 SHs from 190 countries. This means that shareholders from all around the world hold shares in TNCs located in a more restricted number of countries, which, in turn, further concentrates their ownership shares in PCs in an even smaller number of countries, mainly Europe and the US.

In addition, a sector analysis of the LCC shows that the most represented industries are the business activities sector, with 130587 companies, followed by the services

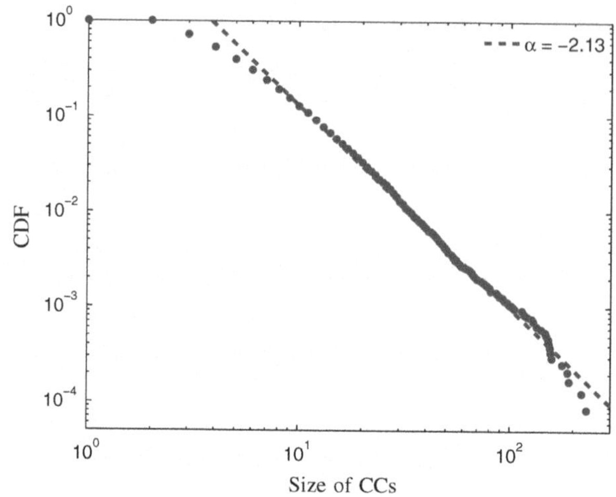


**Fig. 4.4** Cumulative distribution function of the size of the connected components (CCs); the data point representing the largest connected component (LCC), containing 463006 nodes, is not shown, as it is three orders of magnitude larger than second largest (with 230 nodes) and completely offset; as a comparison, a power-law with exponent $-3.13$ ($= \alpha - 1$) is shown

sector with 99839 companies and the manufacturing sector with 66212 companies. On the other hand, surprisingly, the financial intermediaries sector counts only 46632 companies. However, if we distinguish between in-going and out-going relations, the financial intermediaries are the most prevalent shareholders, i.e., represent the companies with the most out-going links.

### 4.2.5 *Strongly Connected Components*

In Sect. 1.2.2 it was pointed out that SCCs in ownership networks correspond to cross-shareholdings of companies. Figure 1.5 on p. 12 shows three examples of different cross-shareholdings.

Recall also from Sect. 1.2.2 that in economics, this kind of ownership relation has raised the attention of different economic institutions, such as the antitrust regulators (which have to guarantee competition in the markets), as well as that of the companies themselves. They can set up cross-shareholdings for coping with possible takeovers, directly sharing information, monitoring and strategies reducing market competition.

In our sample we observe 2219 direct cross-shareholdings (with 4438 ownership relations), in which 2303 companies are involved and represent 0.44 % of all the own-

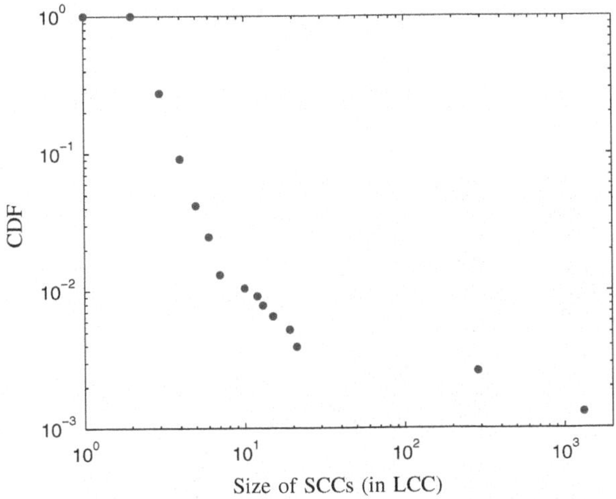


**Fig. 4.5** Cumulative distribution function of the size of the strongly connected components (SCCs) located in the largest connected component (LCC); there exists one dominant SCC

ership relations (see Fig. 1.5a). These direct cross-shareholdings are divided among the different network actors as follow:

- 861 between TNCs;
- 563 between TNCs and PCs;
- 717 between PCs;
- 78 between SHs.

When there is a cross-shareholding involving three companies (see an example in Fig. 1.5b), many combinations of indirect paths are possible. In our network we observe the following ones:

- 829 of the type: $A \rightarrow B \rightarrow C \rightarrow A$;
- 4.395 of the type: $A \leftrightarrow B \rightarrow C \rightarrow A$;
- 8.963 of the type: $A \leftrightarrow B \leftrightarrow C \rightarrow A$;
- 3.129 of the type: $A \leftrightarrow B \leftrightarrow C \leftrightarrow A$.

Next to these simple examples, we also find many SCCs with bigger sizes. Note that smaller SCCs can be embedded in bigger ones. For instance, in the SCC in Fig. 1.5c there is also one cross-shareholding between the nodes $C_I$ and $C_G$. In total there are 915 unique SCCs, of which almost all (83.7 %) are located in the LCC. Focussing only on the LCC, there is one dominant SCC: it is comprised of 1318 companies in 26 countries, see Fig. 4.5. The next smallest SCC contains 286 companies. This is a group of Taiwanese firms located in the OUT of the bow-tie defined by the biggest SCC. The remaining 99.7 % of SCCs in the LCC have sizes between two and 21. The biggest SCC outside the LCC contains 19 firms. In Fig. 4.6 a graph layout of the main SCC is shown.

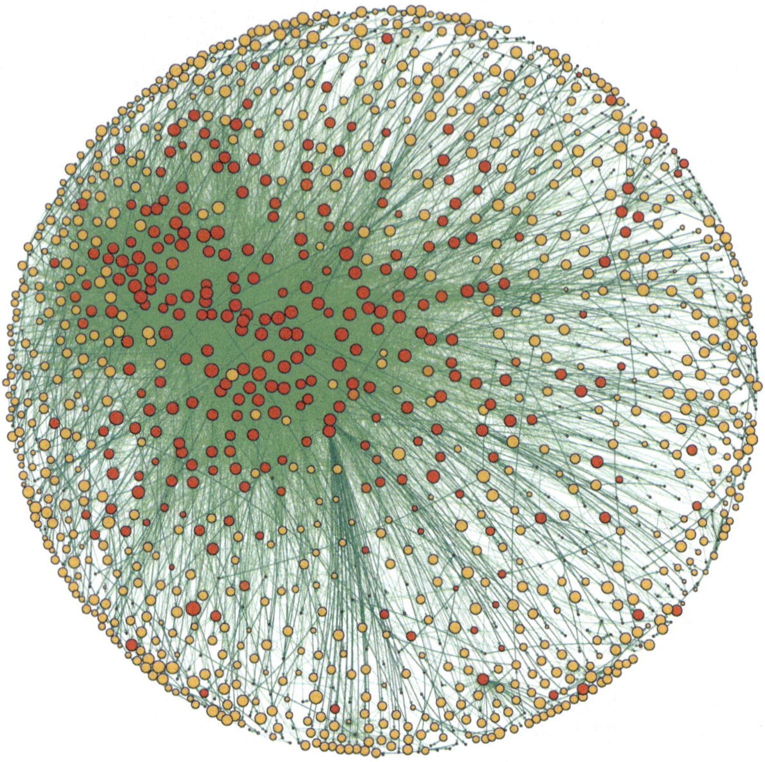

**Fig. 4.6** Network layout (Geipel, 2007) of the SCC (1318 nodes and 12191 links); the node size corresponds to the logarithm of operation revenue, the node color to economic control (from *yellow* to *red*) see details in Sect. 4.3.2; the links are colored (from *light to dark green*) and scaled by weight

The presence of this major SCC may hint at the existence of a rich club (Colizza et al. 2006; Fagiolo et al. 2009). However, the rich club indices proposed in the literature are based on node degree and are thus not suitable for ownership networks, in which indirect and weighted paths matter. Moreover, a reshuffling of the links, necessary to benchmark the result, would lead to economically inviable ownership networks. For instance, exchanging a 10 % ownership share in a small company with 10 % in a big one requires the modification of the budget of the owner. In addition, the procedure is computationally cumbersome for large datasets.

### *4.2.6 The Emerging Bow-Tie Structure*

Having identified one major SCC, we can now unambiguously understand the LCC as a bow-tie structure with this SCC acting as its core. In the following, we only refer

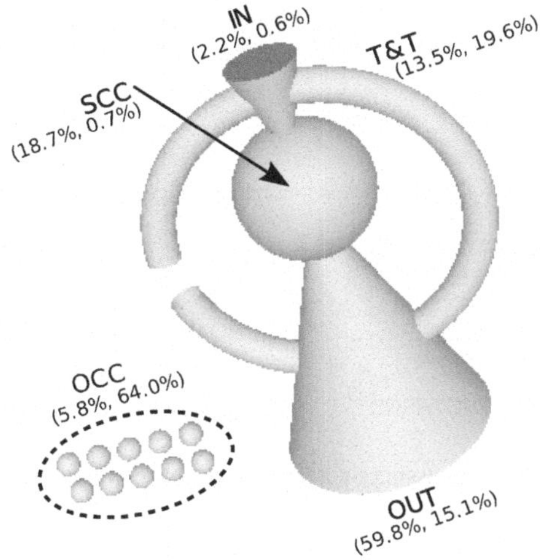


**Fig. 4.7** Illustration of the LCC network bow-tie structure, plus the remaining small OCCs; the volume of the components are scaled by their logarithmic share of TNC operating revenue, seen in the first number below each bow-tie component; the second number corresponds to the percentage of TNCs the component contains; see Table 4.1 on p. 106 for details; the resulting distribution of control is seen in Fig. 4.13

to this SCC. The OUT is significantly larger than the in-section IN and the T&T. The core is also small in comparison. The TNC operating revenue is mainly located in the OUT of the bow-tie, while the SCC, with only 295 TNC, contains about 19 % of the total (see Table 4.1 on p. 106). A diagram representing the whole bow-tie structure of the LCC is shown in Fig. 4.7. Also recall Figs. 1.2 and 3.1. Statistics on the size of the bow-tie components are summarized in Table 4.1.

Does such a bow-tie structure and the relative size of its IN, OUT and SCC result from specific economic mechanisms, or could it be explained by a random network formation process? For correlated networks, as in our case, there is no suitable theoretical prediction (Dorogovtsev et al. 2001). Heuristically, one could address the issue by performing a random reshuffling of links. However, as mentioned, this would violate economic constraints. This issue will be further analyzed in Chap. 5 where we address the challenge of modeling such networks based on growth and link-formation rules. In detail, in Sect. 5.1 we compute the bow-tie sizes of random networks.

Regardless of the sizes of the bow-tie components, what is most important is the existence of a large and dense SCC, which will turn out to play an important role also for control. Indeed, on average a firm in the SCC is owned by 9 other firms of the SCC. Although the weight of individual links is small, 3/4 of the direct ownership of

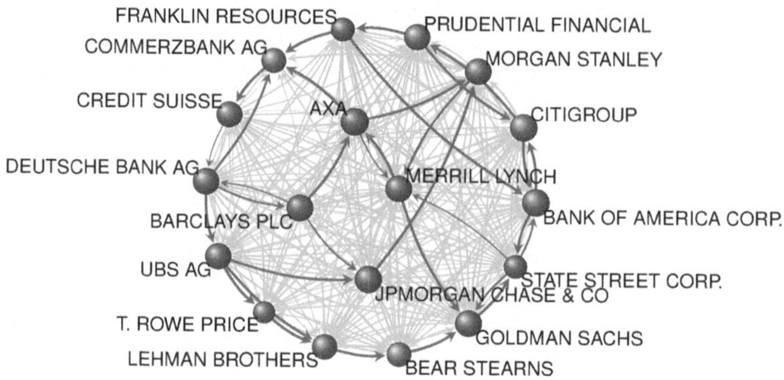


**Fig. 4.8** A subgraph layout of focussing on a few major TNCs in the financial sector; some of the many existing cycles are highlighted

the SCC stays within the SCC itself. In other words, the core consists of a tightly-knit group of corporations that cumulatively hold the majority of each other.

Remarkably, the core is dominated by financial intermediaries which account for 72 % of its nodes. They are mainly located in the US and GB and are related by many mutual cross-shareholdings, as well as longer cycles as shown in Fig. 4.8. One potential consequence of such a situation is the weakening of market competition (Gilo et al. 2006; O'Brien and Salop 1999). In addition, the strong interdependence may also lead to financial instability (Battiston et al. 2007; Lorenz and Battiston 2008). More details of these implications are given in Sect. 6.2.

### *4.2.7 Community Analysis*

In order to further investigate the connectivity of the LCC, we study its community structure. The notion of communities in complex networks is introduced in Appendix B.5. The existence of communities has been investigated in many social networks (Arenas et al. 2004). In ownership networks, economic actors can also organize in communities where they share knowledge, information, capital and control with member companies.

We study the community structure by applying a method that optimizes the modularity (Blondel et al. 2008). This results in the identification of roughly 7000 communities. Most of these are dominated by companies located in a specific geographical region. In particular, there are two major communities, together accounting for about 1/5 of all the nodes and having companies mainly located in the US and GB, respectively (see Table 4.1). Figure 4.9 shows an illustration of the network of the eight biggest communities and their connections. In addition, we also analyze

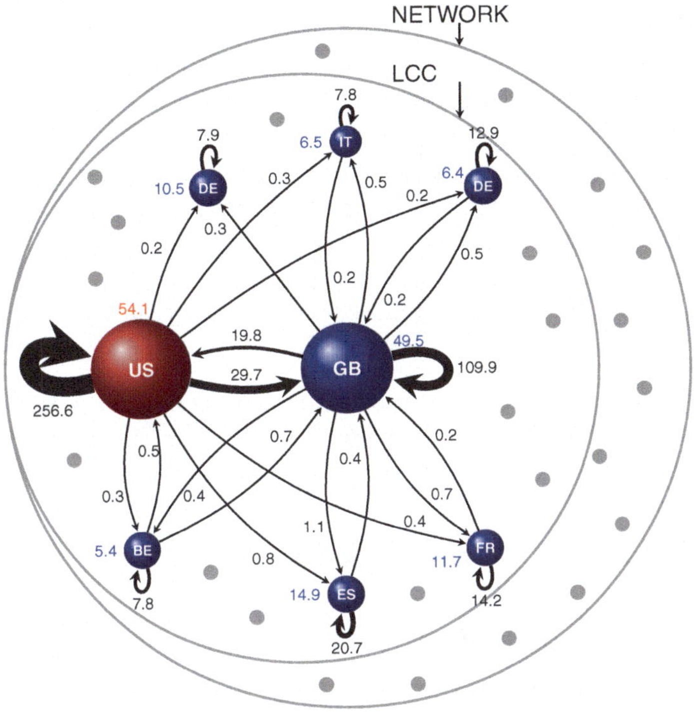


**Fig. 4.9** Illustration of the network of the biggest 8 communities which together have more than 1/3 of all the nodes of the LCC; the colors represent the country dominance of the communities: *red* for the US and *blue* for the European countries; the dimension of the *spheres* and the *arrows* reflects the (logarithmic) size of the communities and the sum of their directed relations, respectively; the numbers indicate the number of links between communities in thousand; the links with less than 50 ownership relations have been omitted

the sub-communities within the biggest communities. The two largest ones display a star-like structure with one large sub-community acting as a hub. Interestingly, the two major communities vertically partition the large bow-tie, having themselves a bow-tie structure (as can be seen in Fig. 4.13).

A more in depth analysis of the role of the financial intermediaries reveals that, although they represent only a small fraction of the economic actors in our sample (9 %), they hold 36 % of all the ownership relations in the network. Many of these relations appear to have a strategic nature. Indeed, the financial intermediaries hold shares larger than 5 % in 13.31 % of non-financial companies and in 60.45 % of other financial companies. Moreover, these intermediaries act as the main connectors between communities. We can verify this by excluding them from the sample. By removing these actors and their relations from the community network, the number of links among communities decreases much more than those within them. Since communities are geographically organized, we can conclude that the financial sectors of different countries play the role of bridging national borders.

### 4.2.8 Summary

**Table 4.1** Summary of the ownership network topology in relation to the economic actor type (SH, TNC and P, in absolute numbers), the TNC operating revenue (OR) in percent and the community analysis (COMM, number of actors per community)

| | TNC (#) | SH (#) | PC (#) | OR (%) | COMM.1 (#) | COMM.2 (#) | COMM.3 (#) |
|---|---|---|---|---|---|---|---|
| LCC | 15491 | 47819 | 399696 | 94.17 | 54065 | 49475 | 14917 |
| IN | 282 | 5205 | 129 | 2.18 | 1355 | 1708 | 940 |
| SCC | 295 | 0 | 1023 | 18.68 | 330 | 356 | 121 |
| OUT | 6488 | 0 | 318073 | 59.85 | 46406 | 39190 | 8955 |
| T&T | 8426 | 42614 | 80471 | 13.46 | 5974 | 8221 | 4901 |
| OCC | 27569 | 29637 | 80296 | 5.83 | - | - | - |

The parts of the network structure analyzed are the following: LCC (IN, OUT, SCC, and T&T) plus the rest of the network, the OCC; in total, there are 43060 TNCs, 600508 economic actors and 1006987 ownership relations

## 4.3 The Network of Global Corporate Control

### 4.3.1 Economics Embedding

Who holds the control over transnational corporations (TNCs)? How much of it is held by the financial sector? How is control achieved through the global network of ownership relations? In the debate on the first question, the idea that corporations are widely held in Anglo-Saxon countries (Berle and Means 1932) has dominated for decades, but has been challenged by recent empirical studies (La Porta et al. 1999; Barca and Becht 2001).[3] Another debate concerns the second question. In reaction to the rise of the so-called "finance capitalism" (Davis 2008), some countries have initially limited the control of financial institutions by means of political interventions (e.g., the Steagal-Glass Act in the US in 1933). More recently, there has been a wave of liberalization of financial markets, motivated by the idea that democratic corporate governance, together with diluted shareholding, can prevent excessive concentration of control (Davis 2008).[4] Regarding the third question, it is known that shareholders wield control over companies not only through direct links, but also through longer indirect pathways. Accordingly, some works have focused on the separation of ownership and control, looking in particular at ownership motifs (e.g., pyramids and cross-shareholdings, see Sect. 1.2.2 on p. 10) (Brioschi et al. 1989; Almeida and Wolfenzon 2006). However, their approach has overlooked the structure of the ownership network as a whole.

In the context of these discussions, our aim is to unveil the global structure of control over the economic value of TNCs. We achieve this by introducing the notion of

[3] See also the discussion given in Sect. 1.2.2 on p. 10.

[4] Again, recall the discussion in Sect. 1.2.2 on p. 10.

network control and integrated control which improves on previous economic models (Brioschi et al. 1989), introduced in Sect. 4.3.2. This quantity estimates the control gained through the network of ownership relations by taking into account all paths among the nodes. A pattern of particular interest for control is the cross-shareholding relation, denoting two firms owning shares of each other (Flath 1992; Dietzenbacher and Temurshoev 2008) (see also Fig. 1.5 on p. 12). A generalization of this kind of relation consists of a group of firms connected by cycles of ownership, which makes the estimation of control non-trivial. Such patterns may have several explanations including: anti-takeover strategies, reduction of transaction costs, risk sharing, increasing trust, existence of groups of interest (Williamson 1975). Whatever their cause, they have implications for market competition (O'Brien and Salop 1999; Gilo et al. 2006).

Despite the importance of TNCs for the global economy, previous investigations in the field of corporate governance have limited their scope to small networks in individual countries (La Porta et al. 1999) and have not investigated the concentration of control. In contrast, our investigation of control deals with the large-scale international sample described in Sect. 4.2.1, looking at the network as a whole, by taking advantage of an analysis of connectivity (Sect. 4.2) and centrality, given in Sect. 4.3.2.

Along these lines, recent work in the field of complex networks has drawn growing attention to economic networks (Schweitzer et al. 2009), including trade (Garlaschelli and Loffredo 2004; Fagiolo et al. 2008), similarity of products (Hidalgo et al. 2007), credit (Boss et al. 2004; Iori et al. 2008), stock price correlation (Bonanno et al. 2003) and shared board directors (Strogatz 2001; Battiston and Catanzaro 2004). This stream of literature has also dealt with ownership (Kogut and Walker 2001; Garlaschelli et al. 2005). In particular, (Glattfelder and Battiston 2009), see Chap. 3, has compared the network of control in different national stock markets. However, in these works, the network is constructed by taking only shareholders at one degree of separation away from companies, thus neglecting many of the indirect ownership relations.

From a theoretical point of view, economics does not offer models that predict how the structure of the global control network should look like. Although it is intuitive that every large corporation has a pyramid of subsidiaries below and a number of shareholders above, it is not clear how much the network surrounding one corporation should be expected to interact with the networks surrounding other corporations. Three alternative hypotheses can be formulated. TNCs may remain isolated, cluster in separated coalitions, or form a giant connected component, possibly with a core-periphery structure. As all of these structures would have different implications for the distribution of control, in a first step, the topological analysis of Sect. 4.2 is imperative in order to uncover the true organization of the market.

### 4.3.2 Level 3: The Flow of Control

The dissection of the network into the various components of the bow-tie structure is not only of interest *per se*, but also necessary for the computation of control. In order to provide an intuitive understanding of the method, we proceed in steps starting from ownership. For a given shareholder $i$, with shares in firms $j$, the value of its portfolio is $\sum_j W_{ij} v_j$, where $v_j$ is the intrinsic value of firms $j$, proxied here by operating revenue, (see also Eq. (1.2)). However, in a network of ownership where the firms $j$, in turn, have shares in other firms $k$, one also needs to account for the value of the portfolios of $j$. Thus, we defined the integrated value $\tilde{v}_i^{\text{int}}$ in Sect. 2.3.1 as the value gained indirectly plus the value of the direct portfolio. In other words: $\tilde{v}_i^{\text{int}} = \sum_j W_{ij} \tilde{v}_j^{\text{int}} + \sum_j W_{ij} v_j$, cf. Eq. (2.44). This notion is also reminiscent of network centrality measures used in sociology and economics (Bonacich 1987; Ballester et al. 2006), in which the score of a node depends recursively on the score of its neighbors. For more details consult Sect. 2.6.

There is also an instructive analogy to a physical system: the direct portfolio value flows upstream in the network (in discrete time steps) from each firm to its shareholders proportionally to the shares they have. The integrated value of a node corresponds then to the total in-flow of value entering the node (Glattfelder and Battiston 2009). See also Sect. 2.4.

To move from ownership to control, we have to estimate how much direct control a shareholder can achieve by owning shares directly in a company. Notice that while ownership is objective, control can only be estimated. This is usually done with a threshold model (TM) explained in Sect. 2.7: if a shareholder holds more than 50 % of ownership, it gets 100 % of control, otherwise control is proportional to the share of ownership, see Fig. 2.5 on p. 53. As a robustness check, we compare the results to a conservative estimate using a linear model (LM), applying the one-share-one-vote rule, and an intermediate, non-linear, relative majority model (RM), introduced in Glattfelder and Battiston (2009). Details can be found in Sect. 2.8.

With these models, the direct control matrix $C$ can be obtained from the ownership matrix $W$, symbolically shown in Eq. (2.116). Now the quantity $\sum_j C_{ij} v_j$ measures the value of direct control shareholder $i$ obtains from its portfolio, introduced in Eq. (2.119). For example, applying the TM model, $W_{ij} = 0.51$ yields $C_{ij} = 1$ and $C_{kj} = 0$, for $k \neq i$. This means, that although shareholder $i$ only holds half of the shares of $j$, it has full control over $j$'s entire intrinsic value. In contrast, for the LM, $C_{ij} = 0.51$, and for the RM, the computation of $C_{ij}$ depends also on the distribution of all the other shares in $j$.

Having obtained the direct control matrix $C$, in a next step, the effect of all indirect paths needs to be accounted for. To achieve this, similarly to the integrated value, we introduced the integrated control in Eq. (2.126) of Sect. 2.9. This quantity measures the value controlled by a shareholder considering the network control of the firms in which it has shares. In matrix notation, $\tilde{\varsigma}^{\text{int}} = \mathcal{C}\tilde{\varsigma}^{\text{int}} + \mathcal{C}v$, which yields $\tilde{\varsigma}^{\text{int}} = (I - \mathcal{C})^{-1}\mathcal{C}v$.

This definition, as it is, faces two problems in the case of an ownership network. Because control flows several times along a cycle, the network control of nodes in any SCC gets overestimated. For the same reason, the network control of nodes in the IN is also overestimated. The first problem has been previously addressed in Baldone et al. (1998) (see details in Sect. 2.3.4), while the second one has been identified and solved here for the first time (details in Sect. 2.3.5).

As described in Sect. 2.5.2, we overcome both issues by developing an algorithm which has to be applied separately to the different components of a bow-tie. As a result, this procedure corrects the definition of network control given above. Furthermore, the computation can be performed for large datasets. Once the corrected network control is determined, denoted by $\overset{\circ}{c}^{\text{net}}$, the integrated control $\overset{\circ}{\varsigma}^{\text{int}}$ is computed as $\overset{\circ}{\varsigma}^{\text{int}} = \overset{\circ}{c}^{\text{net}} - v$, as seen in Eq. (2.130). $\overset{\circ}{c}^{\text{net}}$ is an estimate of the overall value a corporation can control in the network. Notice that network control and integrated control of a company can differ considerably. As an example, Wall Mart is in top rank by operating revenue, hence giving it large network control, but it has no equity share in other TNCs and thus its integrated control is zero. In contrast, a small firm can acquire enormous integrated control via shares in corporations with large operating revenue.

To summarize, using one of the three adjacency matrices estimating direct control, $\mathcal{C}^{LM}$, $\mathcal{C}^{TM}$ and $\mathcal{C}^{RM}$, defined in Eqs. (2.117), (2.118) and (2.125), respectively, we can compute the corresponding network control for a corporation by employing the algorithmic method detailed in Sect. 2.5.2. The resulting network control considers the corrections for the presents of cycles and remedies the problem of root-node accumulation. By deducting the operating revenue $v_i$, we retrieve the corrected integrated control $\overset{\circ}{\varsigma}_i^{\text{int}}$. Recall that the corresponding measures for economic value were $\overset{\circ}{v}_i^{\text{net}}$ of Eq. (2.89) and $\overset{\circ}{v}_i^{\text{int}}$ of Eq. (2.90).

### 4.3.3 Uncovering the Concentration of Control

How much does the network structure matter for network control or integrated control? To answer this question we compute the leverage gained through the network, i.e., the ratio of the integrated control $\overset{\circ}{\varsigma}^{\text{int}}$ over the direct control $c_i$, introduced as control value in Eq. (2.119) of Sect. 2.7. Figure 4.10 displays the cumulative distribution of the leverage, which can be approximated by a power law for a large section of the range. This means that the probability to find actors able to gain a very high network control, compared to the majority, is not negligible and much higher than in a Gaussian distribution. Usually, high leverage is gained by actors having direct shares in small firms, which in turn control big corporations. Indeed, some of them have several levels of indirect ownership below, thus gaining a very high leverage. This

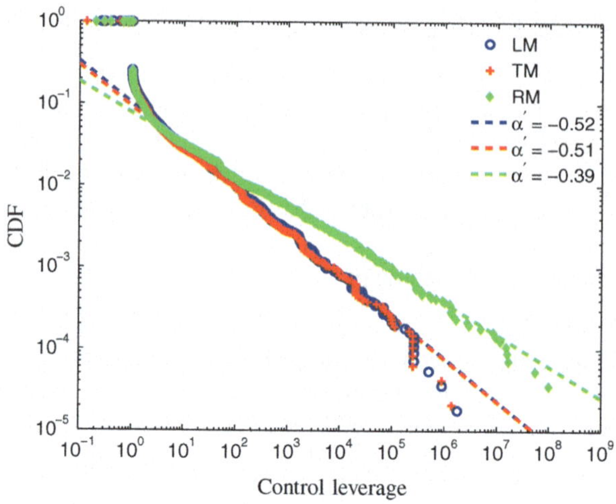


**Fig. 4.10** Cumulative distribution of leverage, which is the ratio between integrated control $\overset{\circ}{\varsigma}^{\text{int}}$ (given by Eq. (2.90) using a control matrix) and the direct control (defined by the control value $c_i$ seen in Eq. (2.119)) for the TNCs and SHs; the *three lines* represent the different models for computing control; the *dashed lines* shows a power-law approximation (CDF $\propto x^{\alpha'}$) to the range of leverage larger than two; the power-law exponents ($\alpha = \alpha' - 1$) for the probability density functions are $-1.52$, $-1.51$, $-1.39$ for LM, TM, RM, respectively; the diagram is in log-log scale

means that the network significantly amplifies the control held by some economic actors.

As the computation of the network control employs the intrinsic value of companies, $\overset{\circ}{c}^{\text{net}} = \overset{\circ}{\varsigma}^{\text{int}} + v$ as see correspondingly in Eq. (2.130), we first look at the distribution of operating revenue. In Fig. 4.11 the various distributions are shown for the LCC. Interestingly, $\overset{\circ}{c}^{\text{net}}$, $\overset{\circ}{\varsigma}^{\text{int}}$ and $v$ share similar functional forms of their PDFs and CDFs which can be approximated by a log-normal distribution. For the TNC operating revenue one finds

$$\mathcal{P}_{\text{op-rev}}(X) = 0.08071 \cdot e^{-\frac{(\ln X - 9.103)^2}{3.36^2}}. \tag{4.1}$$

The goodness-of-fit is given by a value $R^2 = 0.9552$. By visual inspection, the fit deteriorates for very small and very large values. Integrated control and network control are very similar and a fit of $\overset{\circ}{\varsigma}^{\text{net}}$ yields

$$\mathcal{P}_{\text{int-cont}}(X) = 0.1372 \cdot e^{-\frac{(\ln X - 8.558)^2}{4.806^2}}. \tag{4.2}$$

$R^2$ is found to be 0.9949.

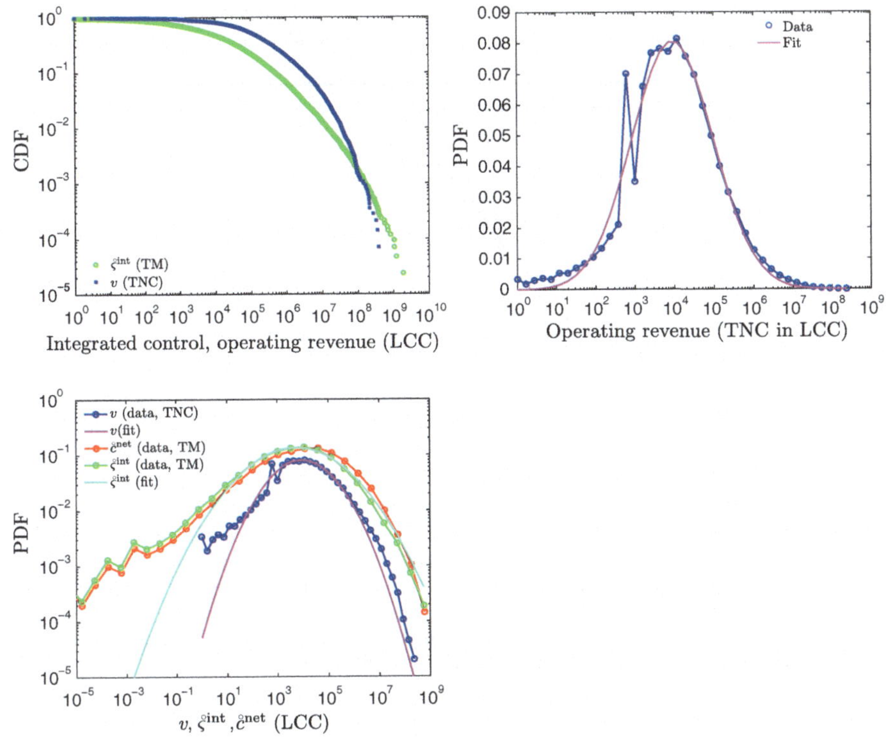


**Fig. 4.11** Probability distributions of operating revenue, integrated control and network control in the LCC; (*top left panel*) comparison of the CDFs of TNC operating revenue and integrated control $\overset{\circ}{\varsigma}^{\text{int}}$ (log-log plot); (*top right panel*) semi-log PDF of the TNC operating revenue $v$, a log-normal curve is fitted over the empirical data, see Eq. (4.1); (*bottom panel*) PDF summarizing the distributions of $v$ for TNC, $\overset{\circ}{\varsigma}^{\text{int}}$ and $\overset{\circ}{c}^{\text{net}}$ (log-log plot), two log-normal fits are shown, see Eqs. (4.1) and (4.2)

However, *a priori*, it is not clear to what degree these distributions result in the concentration of wealth and control. Since many corporations are listed in public stock markets and their shares are accessible to households, one could expect that the control over their economic value is relatively diluted, or, at least, it may be comparable to the inequality of the income distribution across households and firms. On the other hand, because of indirect paths and possible connections among shareholders above the level of the TNCs, one could instead expect that the control over the companies' intrinsic value is more concentrated than this value itself. In order to measure the concentration of network control, we construct a Lorenz-like curve. This procedure is discussed in Sect. 2.10 and is summarized in the caption of Fig. 4.12 and allows us to identify the fraction $\eta^*$ of top power-holders holding together 80 % of the total network control. Thus, the smaller this fraction, the higher the concentration. In our sample we find a value of $\eta_1^* = 0.61\,\%$, corresponding to 737 top holders (TNCs and SHs using the TM). In Table E.1 of Appendix E we provide the list of the first

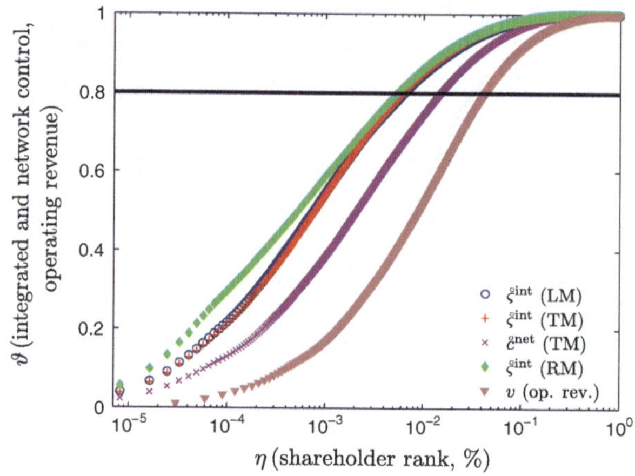


**Fig. 4.12** Concentration of integrated control $\overset{\circ}{\varsigma}^{\text{int}}$, network control $\overset{\circ}{c}^{\text{net}}$ and operating revenue $\upsilon$; economic actors (SHs and TNCs) are sorted by descending importance, as given for instance by their integrated control; a data point located at $(\eta, \vartheta)$ corresponds to a fraction $\eta$ of top economic actors cumulatively holding the fraction $\vartheta$ of integrated control, network control or operating revenue; the different curves refer to integrated control computed with the three models (LM, TM , RM), to network control (only with the TM) and operating revenue; the *horizontal line* denotes a value of $\vartheta$ equal to 80 %; the level of concentration is determined by the $\eta$ value of the intersection between each curve and the *horizontal line*; the diagram is in semi-log scale

50 top ranked holders. This is the first time such a ranking based on global corporate control has been performed.

In order to assess how high this concentration is, we would like to compare our result with other benchmarks. Unfortunately, there are no previous statistics to be found in the literature. However, we can also apply the above procedure to the TNC operating revenue, which yields $\eta_2^* = 4.35\,\%$. Thus, network control is almost one order of magnitude more concentrated than operating revenue. As mentioned, we can also compare our concentration to the ones observed for the income distribution in many developed countries, corresponding to values of $\eta_3^*$ lying between 5 % and 10 % (Atkinson and Bourguignon 2000). Another possible comparison is with the value that we found for the revenue of the top US corporations in the 2009 Fortune1000 data: $\eta_4^* \sim 30\,\%$. This means that integrated control is much more unequally distributed than wealth. Notably, as Fig. 4.12 shows, this result is robust with respect to the models used to estimate direct control from ownership (LM, TM, RM). Only for the top ranked actors the RM yields higher concentration than the other models.

### *4.3.4 The Emerging Picture*

On the one hand we have unveiled the topological structure of the global TNC network and found a dominant bow-tie structure in Sect. 4.2.6. On the other hand, we

**Table 4.2** Probability that a randomly chosen economic actor (TNC or SH) belongs to the group of top power-holders (THPs) with respect to its position in the network structure; the first column refers to all top power-holders, the second column to the first 50 THP

| | All THP (%) | First 50 THP (%) |
|---|---|---|
| IN | 6.233 | 0.273 |
| SCC | 49.831 | 11.525 |
| OUT | 0.432 | 0 |
| T&T | 0.413 | 0.002 |
| OCC | 0.016 | 0 |

**Table 4.3** Number of top power-holder's (TPHs) located in the network structure (SCC and the two biggest communities COMM) and being members of the financial sector (FS), and various intersections thereof; the columns refer to the three models of integrated control and the TM of network control

| | $\mathring{\varsigma}^{int}$ (LM, #) | $\mathring{\varsigma}^{int}$ (TM, #) | $\mathring{\varsigma}^{int}$ (RM, #) | $\mathring{c}^{net}$ (TM, #) |
|---|---|---|---|---|
| TPH | 763 | 737 | 648 | 1791 |
| TPH∩TNC | 308 | 298 | 259 | 1241 |
| TPH∩TNC∩SCC | 151 | 147 | 122 | 211 |
| TPH∩COMM.1 | 253 | 243 | 193 | 633 |
| TPH∩COMM.2 | 177 | 170 | 131 | 374 |
| TPH∩COMM.1∩SCC | 68 | 67 | 56 | 87 |
| TPH∩COMM.2∩SCC | 54 | 51 | 40 | 71 |
| TPH∩SCC∩FS | 116 | 115 | 92 | 140 |

discovered a highly skew distribution of control in the previous section. Consequently we are now able to identify how many top actors are located in each component of the bow-tie and what share of total control they hold.

The main result is that the most powerful actors tend to belong to the core of the bow-tie. Collectively this core holds a large fraction of the total network control (39.0 %), despite being very small compared to the other components. More precisely, among the 737 top holders, 147 are TNCs belonging to the SCC that gain 38.4 % of the total network control. Recalling that there are 295 TNCs present in the core, this implies that a randomly chosen TNC in the SCC has about 50 % chance of also being among the top holders, compared to 6 % for the IN (see Table 4.2). Moreover, at an individual level, the network control held by a node in the SCC is, on average, 20 times larger than that in the IN, whereas being located in the remaining components gives marginal importance to the nodes (see Table 4.3). This suggest that the position of an actor in the network can significantly amplify its level of control. Recalling that the core is a tightly-knit group of corporations that cumulatively hold the majority of each other, then the top power-holders within the core can be thought of as an economic "super-entity" in the global network of corporations.

**Table 4.4** Concentration of 80 % of integrated control (LM, TM, RM) and network control (TM); the percentages refer to the control held by the top power-holder's (TPHs) according to their location in the network structure (SCC and two biggest communities COMM) and them possibly belonging to the FS, and various intersections thereof

| | $\mathring{\varsigma}^{int}$ (LM, %) | $\mathring{\varsigma}^{int}$ (TM, %) | $\mathring{\varsigma}^{int}$ (RM, %) | $\mathring{c}^{net}$ (TM, %) |
|---|---|---|---|---|
| TPH∩TNC | 54.87 | 54.63 | 52.94 | 63.34 |
| TPH∩TNC∩SCC | 39.54 | 38.37 | 37.29 | 30.37 |
| TPH∩COMM.1 | 50.30 | 48.54 | 48.17 | 40.66 |
| TPH∩COMM.2 | 14.92 | 14.43 | 12.42 | 14.50 |
| TPH∩COMM.1∩SCC | 28.19 | 27.62 | 27.17 | 19.44 |
| TPH∩COMM.2∩SCC | 9.17 | 8.55 | 7.78 | 7.58 |
| TPH∩SCC∩FS | 36.58 | 35.37 | 34.90 | 24.36 |

We can also view the core from two additional and complementary perspectives. The details are given in Tables 4.3 and 4.4 (note that qualitatively, the results also hold for network control). Firstly, the insights gained from the community analysis reveal that most of the network control in the core is held by 118 actors in the two top communities (US and GB). The diagram of Fig. 4.13 visualizes this result. Secondly, it turns out that the financial sector holds almost all of the network control of the SCC. In effect, the economic "super-entity" is comprised of high-control wielding financial intermediaries. Figure 4.14 shows a glimpse of this international superstructure of control. Notice that the very existence of such a small, powerful self-owned and self-controlled group of financial TNCs was unsuspected in the economicsliterature. This raises issues regarding the desirability of such a structure for the efficiency of the market and its impact on inequality.

We can also assess the importance of individual countries in terms of integrated control with respect to the operating revenue of their TNCs. Unsurprisingly, the TNCs among the top power-holders in the US have the highest share of operating revenue, integrated control, and network control. In contrast, while JP is second in the ranking of total operating revenue, it looses its position when considering network control (6th). On the other hand, CH gains the most integrated control compared to its share of operating revenue: from 8th to 5th rank.

## Summary

The SCC allows three extremely powerful groups to be discerned: the TNCs, the financial intermediaries and the members of the first community. This is the first time such information has become available and identifies the smallest sets of inter-

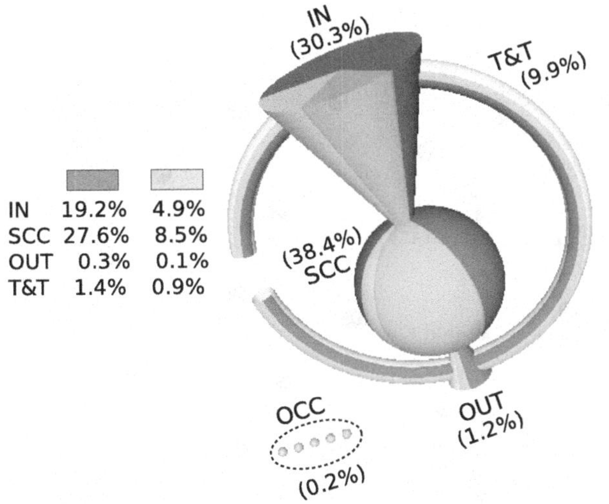


**Fig. 4.13** Illustration of the TNC network featuring the concentration of 80 % of the total integrated control $\mathring{\varsigma}^{net}$; this corresponds to the control held by the 737 top power-holders; the volume of the different components reflects their share of control; the LCC and the remaining network comprised of the OCCs is shown; they contain 79.8 and 0.2 % of integrated control, respectively; the LCC decomposes into a bow-tie structure where the percentage of control is seen in the figure; the *two colored* regions correspond to the two biggest communities, *red* represents the US and *blue* the GB dominated community; the table lists their individual percentages of control; the *largest* part of the integrated control is located in the core of the bow-tie (where 147 top power-holders are TNCs and have 38.4 %) and further concentrated in the biggest community (67 top holders with 27.6 %); the numbers are computed for the TM of control (see Tables 4.3 and 4.4 for the other models); compare this structure with the one seen in Fig. 4.7, reflecting the distribution of (log) TNC operating revenue (We would like to thank D. Garcia for providing the 3D models using processing (http://www.processing.org/.))

connected economic actors able to control disproportionately large fractions of the total integrated control $\mathring{\varsigma}^{int}$. See Table 4.5 for details.

In addition, Table 4.6 shows the top-ten list of the most powerful global economic actors according to their percentage of integrated control. The top-ten power-holders collectively hold 19.45 % of the total integrated control. See also Appendix E.

### 4.3.5 *In Closing*

It is true that countries differ in their legal settings constraining ownership. However, the results of our concentration analysis are robust with respect to three very different models used to infer control from ownership. Moreover, one could argue that we are assigning high control to some financial institutions that may not have interest

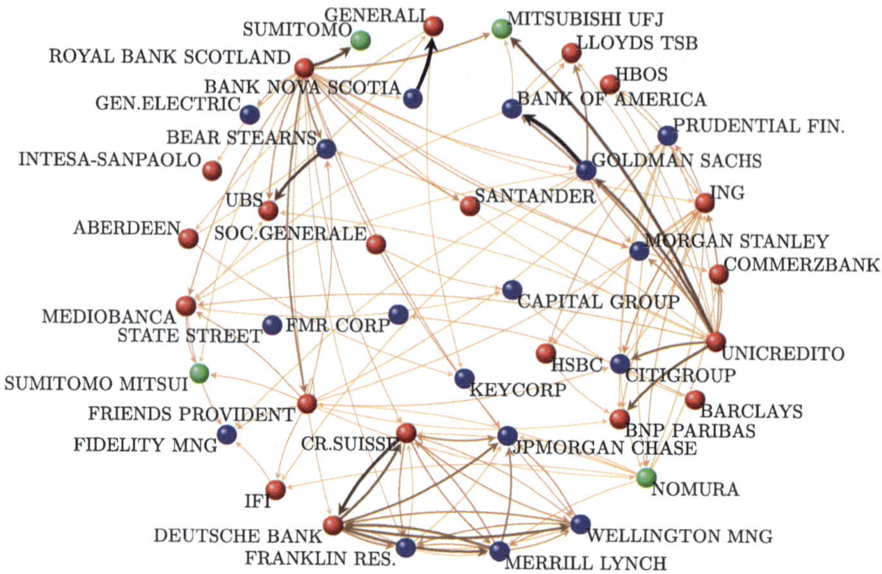


**Fig. 4.14** Sample of the international financial network, where the nodes represent major financial institutions belonging to the SCC and the links give the strongest existing relations among them; *node colors* indicate different geographical areas: EU (*red*), US (*blue*), other countries (*green*); the width and the darkness of the links shows their weight; only the most prominent links are shown; the network shows a high connectivity, with many mutual cross-shareholdings as well as longer cycles; this indicates that the financial sector is strongly interdependent, which make the network vulnerable to instability

in exerting it. This could apply, for instance, to shares held by custodians. Or US mutual funds, when operating in the US, although the same institutions do exert control when operating abroad (Davis 2008). In any case, US mutual funds represent only a small fraction of all global financial institutions. But overall, as there is no empirical evidence to the contrary, we cannot exclude that the top power-holders globally exert their control either formally (e.g., voting in shareholders meetings) or covertly. And in any case, our studies should be seen as the first attempt in uncovering global corporate control. So even approximate or incomplete understanding is still better than no understanding.

Our findings may suggest new questions in different fields of research. From an economics point of view, one could validate empirically the implications on market competition and financial instability. Furthermore, one could systematically investigate the relationship between the profitability of a corporation and its position in the network, and find out whether belonging to the SCC provides a measurable competitive advantage. On the other hand, from the perspective of complex networks, our methodology can be applied to any real-world network with directed and weighted edges to asses the importance of nodes. Moreover, there are many contexts in which

**Table 4.5** The most powerful groups present in the SCC (TNC, FS and COMM.1) and their size (number of nodes and percentage of the 600508 total nodes), the groups total integrated control and the average integrated control each actor holds

| Actor Type | Nodes (#) | Nodes (%) | Total $\mathring{\varsigma}^{int}$ (%) | $\mathring{\varsigma}^{int}$ per Actor (%) |
|---|---|---|---|---|
| TNC | 147 | 0.024 | 38.4 | 0.26 |
| FS | 115 | 0.019 | 35.4 | 0.31 |
| COMM.1 | 67 | 0.011 | 27.6 | 0.41 |

**Table 4.6** List of the first ten corporate top power-holders with country, industrial sector (NACE) and network position information; the list is ordered by integrated control $\mathring{\varsigma}^{int}$ (TM); the top 50 list is given in Table E.1 of Appendix E

| Rank | Economic actor name | Country | NACE code | Network position | Integrated control (TM, %) |
|---|---|---|---|---|---|
| 1 | Barclays Plc | GB | 6512 | SCC | 4.05 |
| 2 | Capital Group Companies Inc, The | US | 6713 | IN | 2.61 |
| 3 | Fmr Corp | US | 6713 | IN | 2.28 |
| 4 | Axa | FR | 6712 | SCC | 2.27 |
| 5 | State Street Corporation | US | 6713 | SCC | 1.81 |
| 6 | Jpmorgan Chase & Co. | US | 6512 | SCC | 1.53 |
| 7 | Legal & General Group Plc | GB | 6603 | SCC | 1.47 |
| 8 | Vanguard Group, Inc., The | US | 7415 | IN | 1.23 |
| 9 | Ubs Ag | CH | 6512 | SCC | 1.21 |
| 10 | Merrill Lynch & Co., Inc. | US | 6712 | SCC | 0.99 |

it is possible to assign a value to the nodes and obtain a more accurate estimate. Finally, our method could help in the development of theoretical models that explain the emergence of bow-tie structures in presence of correlation, see Chap. 5.

For an in-depth discussion of the relevance of our work and a summary of the overall implications, consult Chap. 6.

## References


H.V. Almeida, D. Wolfenzon, A Theory of pyramidal ownership and family business groups. J. Finan. **61**(6), 2637–2680 (2006)

A. Arenas, L. Danon, A. Diaz-Guilera, P.M. Gleiser, R. Guimera, Community analysis in social networks. Eur. Phys. J. B **38**(2), 373–380 (2004)

A.B. Atkinson, F. Bourguignon, *Handbook of Income Distribution* (Elsevier, Amsterdam, 2000)

S. Baldone, F. Brioschi, S. Paleari, Ownership measures among firms connected by cross-shareholdings and a further analogy with input-output theory, in *4th JAFEE International Conference on Investment and Derivatives*, Madison, 1998

C. Ballester, A. Calvo-Armengol, Y. Zenou, Who's who in networks. Wanted: the key player. Econometrica **74**(5), 1403–1417 (2006)

F. Barca, M. Becht, *The Control of Corporate Europe* (Oxford University Press, Oxford, 2001)
S. Battiston, M. Catanzaro, Statistical properties of corporate board and director networks. Eur. Phys. J. B **38**(2), 345–352 (2004)
S. Battiston, D. Delli Gatti, M. Gallegati, B. Greenwald, J.E. Stiglitz, Credit chains and bankruptcy propagation in production networks. J. Econ. Dyn. Control **31**(6), 2061–2084 (2007)
A. Berle, G. Means, *The Modern Corporation and Private Property* (Mac Millan, New York, 1932)
V.D. Blondel, J.L. Guillaume, R. Lambiotte, E.L.J.S. Mech, Fast unfolding of communities in large networks. J. Stat. Mech **2008**, P10008 (2008)
P. Bonacich, Power and centrality: a family of measures. The Am. J. Sociol. **92**(5), 1170–1182 (1987)
G. Bonanno, G. Caldarelli, F. Lillo, R.N. Mantegna, Topology of correlation-based minimal spanning trees in real and model markets. Phys. Rev. E **68**(4), 46130 (2003)
M. Boss, H. Elsinger, M. Summer, S. Thurner, Network topology of the interbank market. Quant. Finan. **4**(6), 677–684 (2004)
F. Brioschi, L. Buzzacchi, M.G. Colombo, Risk capital financing and the separation of ownership and control in business groups. J. Bank. Finance **13**(1), 747–772 (1989)
V. Colizza, A. Flammini, M.A. Serrano, A. Vespignani, Detecting rich-club ordering in complex networks. Nat. Phys **2**(2), 110–115 (2006)
G.F. Davis, A new finance capitalism? mutual funds and ownership re-concentration in the united states. Eur. Manag. Rev. **5**(1), 11–21 (2008)
D. Deutsch, *The Fabric of Reality* (Penguin Books, London, 1998)
E. Dietzenbacher, U. Temurshoev, Ownership relations in the presence of cross-shareholding. J. Econ. **95**(3), 189–212 (2008)
S.N. Dorogovtsev, J.F.F. Mendes, A.N. Samukhin, Giant strongly connected component of directed networks. Phys. Rev. E **64**(2), 25101 (2001)
G. Fagiolo, J. Reyes, S. Schiavo, On the topological properties of the world trade web: a weighted network analysis. Phys. A. **387**, 3868–3873 (2008)
G. Fagiolo, J. Reyes, S. Schiavo, World-trade web: topological properties, dynamics, and evolution. Phys. Rev. E **79**(3), 36115 (2009)
D. Flath, Indirect shareholding within Japan's business groups. Econ. Lett. **38**(2), 223–227 (1992)
D. Garlaschelli, M.I. Loffredo, Structure and evolution of the world trade network. Phys. A **355**, 138–144 (2004)
D. Garlaschelli, S. Battiston, M. Castri, V.D.P. Servedio, G. Caldarelli, The scale-free topology of market investments. Phys. A **350**, 491–499 (2005)
M.M. Geipel, Self-organization applied to dynamic network layout. Int. J. Mod. Phys. C **18**(10), 1537–1549 (2007)
D. Gilo, Y. Moshe, Y. Spiegel, Partial cross ownership and tacit collusion. RAND J. Econ. **37**(1), 81–99 (2006)
J.B. Glattfelder, S. Battiston, Backbone of complex networks of corporations: the flow of control. Phys. Rev. E **80**(3), 36104 (2009)
C.A. Hidalgo, B. Klinger, A.L. Barabási, R. Hausmann, The product space conditions the development of nations. Science **317**(5837), 482 (2007)
G. Iori, G. De Masi, O.V. Precup, G. Gabbi, G. Caldarelli, A network analysis of the italian overnight money market. J. Econ. Dyn. Control **32**(1), 259–278 (2008)
B. Kogut, G. Walker, The small world of germany and the durability of national networks. Am. Sociol. Rev. **66**(3), 317–335 (2001)
R. La Porta, F. Lopez de Silanes, A. Shleifer, Corporate ownership around the world. J. Finan. **54**(2), 471–517 (1999)
J. Lorenz, S. Battiston, Systemic risk in a network fragility model analyzed with probability density evolution of persistent random walks. Netw. Heterogen. Media **3**(2), 185 (2008)
D.P. O'Brien, S.C. Salop, Competitive effects of partial ownership: financial interest and corporate control. Antitrust Law J. **67**, 559–614 (1999)
OECD. The OECD guidelins for multinational enterprises, (2000) www.oecd.org

F. Schweitzer, G. Fagiolo, D. Sornette, F. Vega-Redondo, A. Vespignani, D.R. White, Economic networks: the new challenges. Science **325**(5939), 422 (2009)

S.H. Strogatz, Exploring complex networks. Nature **410**(6825), 268–276 (2001)

S. Vitali, *Industrial Organization from a Geographical and Network Perspective*. Ph.D. Thesis, Chair of Systems Design, ETH, Zurich, 2010

S. Vitali, J.B. Glattfelder, S. Battiston, The network of global corporate control. PloS ONE **6**(10), e25995 (2011)

O.E. Williamson, *Markets and Hierarchies, Analysis and Antitrust Implications: A Study in the Economics of Internal Organization* (Free Press, New York, 1975)

# Chapter 5
# The Bow-Tie Model of Ownership Networks

*Indeed, even some of the very simplest programs that I looked at had behavior that was as complex as anything I had ever seen.*
*It took me more than a decade to come to terms with this result, and to realize just how fundamental and far-reaching its consequences are.*

(S. Wolfram in Wolfram 2002, p. 2)

Perhaps the most surprising feature discovered in the empirical network analysis of Chap. 4 is the emergence of a tiny, powerful, tightly-knit and self-controlled group of corporations, see Sect. 4.3.5. This core can be identified as the strongly connected component (SCC[1]) of an emerging bow-tie structure[2] in the global network of TNCs, located in the largest connected component (LCC) of the network, see Sect. 4.2.4. Collectively this core holds close to 40 % of the total control in the network, despite being comprise of only 1347 corporations. Recall that the network size is 600508. The relevance of this structure is discussed in Sect. 6.1.6 and its implications in Sect. 6.2. The emergence of a bow-tie topology in the global ownership network, has, to our knowledge, never been observed before. Recall also that in the cross-country analysis of Chap. 3 we also uncovered bow-tie structures in various national networks, see Sect. 3.3.1.

It is known that technological networks, such as the World-Wide Web (WWW) (Broder et al. 2000) and Wikipedia[3] (Capocci et al. 2006), also exhibit bow-tie topologies. In contrast to the ownership networks, their SCCs are large, comprising more than half of all the nodes.

What are the organizational principles and the driving forces behind this kind of network organization? In order to understand the mechanisms underlying the formation of different bow-tie structures, we develop a generic Modeling Framework in Sect. 5.2. It is governed by node and link addition, where the network evolution

---

[1] A list of acronyms can be found in front matter of this book.

[2] Recall Figs. 1.2 and 3.1.

[3] The Internet encyclopedia, http://wikipedia.org.

J. B. Glattfelder, *Decoding Complexity*, Springer Theses,
DOI: 10.1007/978-3-642-33424-5_5, © Springer-Verlag Berlin Heidelberg 2013

is determined by a preferential-attachment mechanism defined by a distribution of fitness values amongst the nodes. This fitness measure can either be determined by the network topology (e.g., degree, centrality, network control, etc.) or can be a non-topological state variable (e.g., operating revenue). The network formation is determined by the co-evolution of fitness and topology: at each time step, the distribution of fitness determines the topology of the network, which in turn impacts the distribution of fitness in the next step.

Using the Modeling Framework we can address the question of what the simplest mechanisms are that result in the emergence of bow-tie structures. In other words, what interactions are necessary at the micro-level in order to reproduce the observed macro-patterns. In detail, in Sect. 5.3, we present a specific incarnation of the framework to reproduce the empirical properties of the TNC network. We focus on the specific sizes of the bow-tie components and the degree distribution. This sheds new light on the possible interaction-mechanisms of economic agents in ownership networks.

First, in Sect. 5.1, we discuss some general network-theoretical aspects of bow-tie topologies in networks where the nodes are uncorrelated. These insights can be understood as our null hypothesis: the emergence of bow-tie structures in random networks.

## 5.1 Bow-Tie Components Size of Networks

What structures can be expected in generic, uncorrelated networks: can random networks exhibit bow-tie topologies? Note that the Appendices B.6.1 and B.6.2 cover some details of undirected and directed random graphs, respectively.

### *5.1.1 Theoretical Components Size of Directed Networks*

To summarize, in directed networks, a (weakly) connected component (CC) refers to the set of nodes and vertices that are connected regardless of the direction of the links. Similarly, a strongly SCC is a subgraph in which each node is reachable from every other node by a chain of directed links.

Within any CC there could be many SCCs. By choosing a SCC, it is possible to define a bow-tie topology, with the SCC as its core. Furthermore, there often exists a large connected component next to smaller disconnected components in arbitrary networks. The LCCs largest SCC (LSCC) unambiguously defines the predominant bow-tie structure.

The size of the LCC for an undirected graph can be computed from the following heuristic argument. Let $u$ be the fraction of nodes not in the LCC. The probability that a node $i$ is not in the LCC is equal to the probability that none of its neighbors belong to the LCC, which is given by $u^{k_i}$. Thus the following self-consistency relation holds

$$u = \sum_k \mathcal{P}(k)u^k =: \phi(u). \tag{5.1}$$

Note that the right-hand side of Eq. (5.1) defines a so-called generating function $\phi(u)$ (see (Durrett 2004) for details). The size of the LCC, denoted by $\mathcal{L}$, can be computed from the relations $\mathcal{L} = 1 - \phi(u)$ and $\phi(u) = u$, as will be demonstrated below for the directed case.

Similarly, using the generating function formalism for directed graphs (Newman et al. 2001; Dorogovtsev et al. 2001)

$$\phi(x, y) := \sum_{k_{in},k_{out}} \mathcal{P}(k_{in}, k_{out})x^{k_{in}}y^{k_{out}}, \tag{5.2}$$

the components size can be computed analytically as a function of the link probability $p_{tot} = z/n$, where $z = \langle k_{tot}\rangle$ in the directed case. Note that $\zeta := z/2 = z_{in} = z_{out}$. Furthermore

$$\zeta = \partial_x\, \phi(x, 1)|_{x=1} = \partial_y\, \phi(1, y)|_{y=1}\,. \tag{5.3}$$

Defining

$$\phi_1^{LCC}(x) := \frac{1}{z}\partial_x\phi(x, x), \tag{5.4}$$

$$\phi_1^{IN}(x) := \frac{1}{\zeta}\partial_y\, \phi(x, y)|_{y=1}\,, \tag{5.5}$$

$$\phi_1^{OUT}(y) := \frac{1}{\zeta}\partial_x\, \phi(x, y)|_{x=1}\,, \tag{5.6}$$

the components size obey

$$\mathcal{L} := 1 - \phi(x_c, x_c); \quad x_c = \phi_1^{LCC}(x_c), \tag{5.7}$$

$$\widehat{\mathcal{I}} := 1 - \phi(x_c, 1); \quad x_c = \phi_1^{IN}(x_c), \tag{5.8}$$

$$\widehat{\mathcal{O}} := 1 - \phi(1, y_c); \quad y_c = \phi_1^{OUT}(y_c). \tag{5.9}$$

Note that the values $\widehat{\mathcal{I}}$ and $\widehat{\mathcal{O}}$ both also contain the size of the SCC, denoted by $\mathcal{S}$, which is given by

$$\begin{aligned}\mathcal{S} &:= 1 - \phi(x_c, 1) - \phi(1, y_c) + \phi(x_c, y_c)\\ &= \widehat{\mathcal{I}} \cdot \widehat{\mathcal{O}} - \phi(x_c, 1)\cdot\phi(1, y_c) + \phi(x_c, y_c).\end{aligned} \tag{5.10}$$

This means that the sizes of the IN and OUT components is given by

$$\mathcal{I} := \widehat{\mathcal{I}} - \mathcal{S}, \tag{5.11}$$

$$\mathcal{O} := \widehat{\mathcal{O}} - \mathcal{S}. \tag{5.12}$$

Finally, for the tubes and tendrils (T&T) one finds

$$\mathcal{T} := \mathcal{L} + \mathcal{S} - \widehat{\mathcal{I}} - \widehat{\mathcal{O}} = \mathcal{L} - \mathcal{S} - \mathcal{I} - \mathcal{O}. \tag{5.13}$$

## 5.1.2 Components Size of Directed Random Networks

Applying this methodology to the case of a directed random graph, described in Appendix B.6.2, the generating function is

$$\begin{aligned}\phi(x,y) &= \sum_{k_{in},k_{out}} \mathcal{P}(k_{in},k_{out}) x^{k_{in}} y^{k_{out}} = \sum_{k_{in},k_{out}} \mathcal{P}(k_{in})\mathcal{P}(k_{out}) x^{k_{in}} y^{k_{out}} \\ &= \sum_{k_{in}} \frac{(\zeta x)^{k_{in}} e^{-\zeta}}{k_{in}!} \sum_{k_{out}} \frac{(\zeta y)^{k_{out}} e^{-\zeta}}{k_{out}!} = e^{-z+\zeta(x+y)}. \end{aligned} \tag{5.14}$$

Hence

$$\phi_1^{LCC}(x) = e^{z(x-1)} = \phi(x,x), \tag{5.15}$$

$$\phi_1^{IN}(x) = e^{\zeta(x-1)}, \tag{5.16}$$

$$\phi_1^{OUT}(y) = e^{\zeta(y-1)}. \tag{5.17}$$

From Eq. (5.7) the LCC size is found to be

$$\mathcal{L} = 1 - e^{z(x_c-1)}, \tag{5.18}$$

$$x_c = e^{z(x_c-1)}, \tag{5.19}$$

and it holds that $x_c = 1 - \mathcal{L}$, which, substituted into Eq. (5.18), yields

$$\mathcal{L} = 1 - e^{-z\mathcal{L}}. \tag{5.20}$$

Correspondingly, the IN plus SCC size, respectively the OUT plus SCC size are

$$\widehat{\mathcal{I}} = 1 - e^{-\zeta \widehat{\mathcal{I}}}, \tag{5.21}$$

$$\widehat{\mathcal{O}} = 1 - e^{-\zeta \widehat{\mathcal{O}}}. \tag{5.22}$$

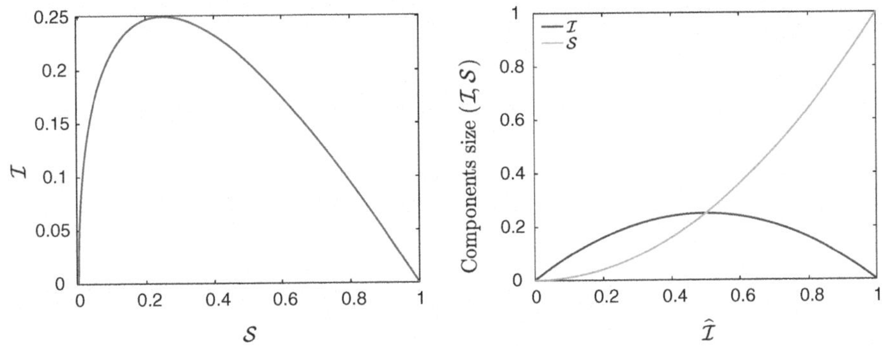


**Fig. 5.1** Bow-tie components size as functions of each other: (*left*) $\mathcal{I}(\mathcal{S})$ from Eq. (5.25); (*right*) $\mathcal{I}$ and $\mathcal{S}$ as functions of $\widehat{\mathcal{I}}$ from Eqs. (5.23) and (5.26), respectively

This reveals that $\widehat{\mathcal{I}} = \widehat{\mathcal{O}}$. Equations. (5.20)–(5.22) are transcendental with no closed-form solutions. They can be numerically solved by computing the zeros of the function $f(\mathcal{C}) = 1 - e^{-z\mathcal{C}} - \mathcal{C}$, $\mathcal{C}$ being the component's size, for varying $z$.

Note that because Eq. (5.14) factorizes (see Appendix B.6.2), Eq. (5.10) yields

$$\mathcal{S} = \widehat{\mathcal{I}}^2 = \widehat{\mathcal{O}}^2. \tag{5.23}$$

Once the size of the SCC is determined, $\mathcal{S} = \mathcal{S}(\widehat{\mathcal{I}}, \widehat{\mathcal{O}})$ the sizes of $\mathcal{I} = \mathcal{I}(\mathcal{S})$ can be computed from Eqs. (5.11) and (5.23) as

$$\mathcal{I}^2 + 2\mathcal{I}\mathcal{S} + \mathcal{S}^2 - \mathcal{S} = 0, \tag{5.24}$$

with the positive solution

$$\mathcal{I} = -\mathcal{S} + \sqrt{\mathcal{S}}. \tag{5.25}$$

The maximum of $\mathcal{I}(\mathcal{S})$ is reached for $\mathcal{S}^* = 0.25$, and $\mathcal{I}(\mathcal{S}^*) = 0.25$. The left-side diagram in Fig. 5.1 shows a plot of Eq. (5.25). It is also straightforward to express $\mathcal{I} = \mathcal{I}(\widehat{\mathcal{I}})$, i.e.,

$$\mathcal{I} = \widehat{\mathcal{I}} - \widehat{\mathcal{I}}^2, \tag{5.26}$$

which is shown in the right-side diagram of Fig. 5.1. The two functions in Eqs. (5.23) and (5.26) are equal if $\widehat{\mathcal{I}} = 0.5$. In other words, $\mathcal{I}(0.5) = \mathcal{S}(0.5) = 0.25$. Note that identical relations hold for $\mathcal{O} = \mathcal{O}(\widehat{\mathcal{O}})$.

In Fig. 5.2 the analytical values are plotted for the various components size as a function of the average degree. In addition, empirical simulation results are shown.

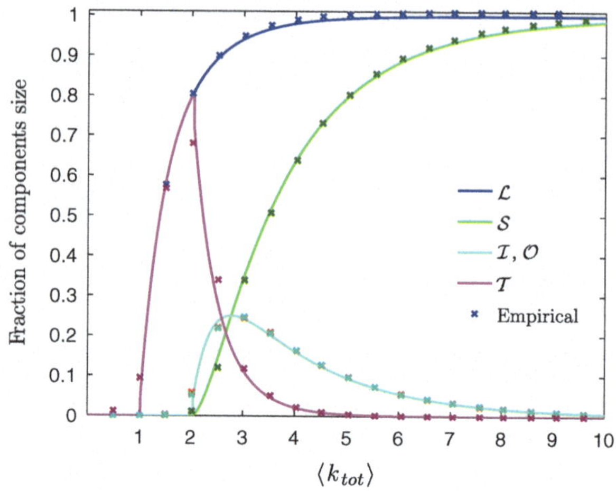


**Fig. 5.2** Fraction of components size for the LCC ($\mathcal{L}$), LSCC ($\mathcal{S}$), IN ($\mathcal{I}$), OUT ($\mathcal{O}$), and T&T ($\mathcal{T}$) as a function of the total average degree $\langle k_{tot} \rangle = z = pn$; each data point denotes the average empirical values gained from 100 simulations of 1000-node networks for fixed $p$

From Eq. (5.21) the corresponding value of $z = 2\zeta = \langle k_{tot} \rangle$ can be computed as

$$z = -\frac{2 \ln(1 - \widehat{\mathcal{I}})}{\widehat{\mathcal{I}}}. \tag{5.27}$$

The maximal size of $\mathcal{I} = 0.25$ is reached at $\widehat{\mathcal{I}} = 0.5$, thus $z^*_{IN} = -4 \ln(0.5) \approx 2.773$. As noted, at this point $\mathcal{I} = \mathcal{S}$.

### 5.1.3 Empirical TNC Components Size

We can now compare this theoretical result with the real-world global ownership network. Figure 5.3 shows the empirical components size of the TNC network for the average total degree $\langle k_{tot} \rangle \approx 3.358$.

In Table 5.1 the empirical values are compared to the theoretical ones of a directed random graph (DRG). It is apparent, that the TNC network heavily deviates from a random network.

To summarize, the following main differences between the empirically observed patterns of the TNC network and directed random graphs are observed:

- the empirical LCC size is smaller ($\mathcal{L}^{TNC} < \mathcal{L}^{DRG}$);
- only the random network has identical IN and OUT sizes, for the TNC: $\mathcal{I}^{TNC} \neq \mathcal{O}^{TNC}$;
- the TNC network's OUT size is very large ($\mathcal{O}^{TNC} > \mathcal{O}^{DRG}_{max} = 0.25$);
- the empirical IN size is very small ($\mathcal{I}^{TNC} < \mathcal{I}^{DRG}$);
- the TNC network's SCC is tiny ($\mathcal{S}^{TNC} \ll \mathcal{S}^{DRG}$);

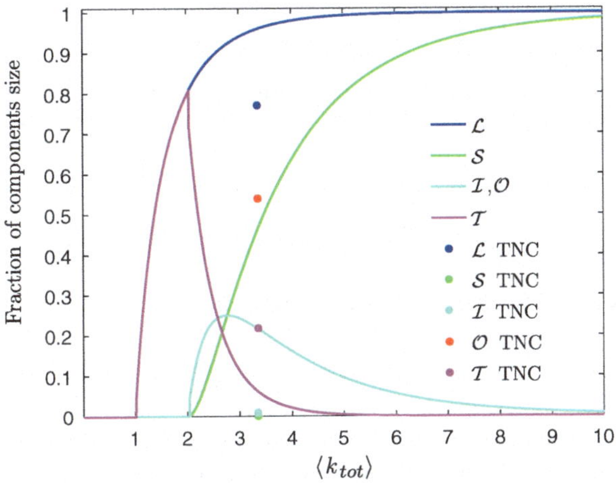


**Fig. 5.3** Comparing the results of Fig. 5.2 with the components size of the TNC network shown as filled disks for the empirical value $\langle k_{tot} \rangle \approx 3.358$ and inferred $p = 5.59 \cdot 10^{-6}$

**Table 5.1** Empirical and theoretical components size; the percentage values are with respect to the total network size (i.e, the LCC plus OCC)

| | $\mathcal{L}$ | $\widehat{\mathcal{I}}$ | $\widehat{\mathcal{O}}$ | $\mathcal{S}$ | $\mathcal{I}$ | $\mathcal{O}$ | $\mathcal{T}$ |
|---|---|---|---|---|---|---|---|
| TNC | 0.7710 | 0.0095 | 0.5406 | 0.0002 | 0.0093 | 0.5404 | 0.2189 |
| DRG | 0.9602 | 0.6815 | 0.6815 | 0.4645 | 0.2170 | 0.2170 | 0.0616 |

- tiny SCC values in the DRG result in the bow-tie being mostly comprised of T&T: $\mathcal{S}^{DRG} \approx 0 \Rightarrow \mathcal{T}^{DRG} \approx \mathcal{L}^{DRG}$.

In effect, there exists no value for $\langle k_{tot}^{DRG} \rangle$ where the empirical and theoretical components size are comparable.

### *5.1.4 Components Size of Generalized Directed Random Networks*

We now consider an extension of directed Poisson random networks that are obtained from a given *arbitrary* degree distributions for the in- and out-degree. It is customary to call such networks *generalized (directed) random networks*.

In the case of uncorrelated generalized random networks Eqs. (5.7)–(5.10) still apply and allow the components size $\mathcal{L}$, $\widehat{\mathcal{I}}$, and $\widehat{\mathcal{O}}$ to be computed. For $\mathcal{S}$, one can define

$$\mathcal{S} = \widehat{\mathcal{I}} \cdot \widehat{\mathcal{O}} + \Delta, \tag{5.28}$$

where $\Delta := -\phi(x_c, 1) \cdot \phi(1, y_c) + \phi(x_c, y_c)$ is non-zero for non-factorizable joint probability densities. Recall Appendix B.6.2.

### Scale-Free Directed Networks

Scale-free networks are introduced in Appendix B.3.1. In the undirected case, their generating function is given as

$$\phi(x) = \sum_k \mathcal{P}(k)x^k = \frac{\mathrm{Li}_\alpha(xe^{-1/\kappa})}{\mathrm{Li}_\alpha(e^{-1/\kappa})}. \tag{5.29}$$

Note that due to the definition of the polylogarithm

$$\frac{d\phi(x)}{dx} = \frac{\mathrm{Li}_{\alpha-1}(xe^{-1/\kappa})}{x\mathrm{Li}_{\alpha-1}(e^{-1/\kappa})}. \tag{5.30}$$

In the directed case, it follows that the generating function takes the form

$$\begin{aligned}
\phi(x,y) &= \sum_{k_{in},k_{out}} \mathcal{P}(k_{in})P(k_{out})x^{k_{in}}y^{k_{out}} && (5.31)\\
&= \sum_{k_{in}} \frac{k_{in}^{-\alpha}e^{-k_{in}/\kappa_{in}}}{\mathrm{Li}_\alpha(e^{-1/\kappa_{in}})}x^{k_{in}} \sum_{k_{out}} \frac{k_{out}^{-\beta}e^{-k_{out}/\kappa_{out}}}{\mathrm{Li}_\beta(e^{-1/\kappa_{out}})}y^{k_{out}} && (5.32)\\
&= \frac{\mathrm{Li}_\alpha(xe^{-1/\kappa_{in}})}{\mathrm{Li}_\alpha(e^{-1/\kappa_{in}})}\frac{\mathrm{Li}_\beta(ye^{-1/\kappa_{out}})}{\mathrm{Li}_\beta(e^{-1/\kappa_{out}})}. && (5.33)
\end{aligned}$$

As a first result, the average degree can be computed from Eq. (5.3)

$$\zeta = \frac{\mathrm{Li}_{\alpha-1}(e^{-1/\kappa_{in}})}{\mathrm{Li}_\alpha(e^{-1/\kappa_{in}})} = \frac{\mathrm{Li}_{\beta-1}(e^{-1/\kappa_{out}})}{\mathrm{Li}_\beta(e^{-1/\kappa_{out}})}. \tag{5.34}$$

The minimum $\zeta$ is reached for $\alpha \to \infty$ ($\mathrm{Li}_\alpha(x) \to x$) and $\kappa \to 0$. This implies that $\zeta \geq 1$ and $z \geq 2$.

Equation (5.4) now reads

$$\phi_1^{LCC}(x) = \frac{\mathrm{Li}_{\alpha-1}(xe^{-1/\kappa_{in}})\mathrm{Li}_\beta(xe^{-1/\kappa_{out}}) + \mathrm{Li}_\alpha(xe^{-1/\kappa_{in}})\mathrm{Li}_{\beta-1}(xe^{-1/\kappa_{out}})}{zx\mathrm{Li}_\alpha(e^{-1/\kappa_{in}})\mathrm{Li}_\beta(e^{-1/\kappa_{out}})}. \tag{5.35}$$

The size of the LCC is given by Eq. (5.7)

$$\mathcal{L} = 1 - \frac{\mathrm{Li}_\alpha(x_c e^{-1/\kappa_{in}})}{\mathrm{Li}_\alpha(e^{-1/\kappa_{in}})}\frac{\mathrm{Li}_\beta(x_c e^{-1/\kappa_{out}})}{\mathrm{Li}_\beta(e^{-1/\kappa_{out}})}; \quad x_c = \phi_1^{LCC}(x_c). \tag{5.36}$$

Correspondingly, for $\widehat{\mathcal{I}}$ and $\widehat{\mathcal{O}}$

$$\phi_1^{IN}(x) = \frac{\mathrm{Li}_\alpha(xe^{-1/\kappa_{in}})}{\zeta \mathrm{Li}_\alpha(e^{-1/\kappa_{in}})} \frac{\mathrm{Li}_{\beta-1}(e^{-1/\kappa_{out}})}{\mathrm{Li}_\beta(e^{-1/\kappa_{out}})}, \tag{5.37}$$

$$\phi_1^{OUT}(x) = \frac{\mathrm{Li}_{\alpha-1}(e^{-1/\kappa_{in}})}{\zeta \mathrm{Li}_\alpha(e^{-1/\kappa_{in}})} \frac{\mathrm{Li}_\beta(ye^{-1/\kappa_{out}})}{\mathrm{Li}_\beta(e^{-1/\kappa_{out}})}, \tag{5.38}$$

and

$$\widehat{\mathcal{I}} = 1 - \frac{\mathrm{Li}_\alpha(x_c e^{-1/\kappa_{in}})}{\mathrm{Li}_\alpha(e^{-1/\kappa_{in}})}; \quad x_c = \phi_1^{IN}(x_c), \tag{5.39}$$

$$\widehat{\mathcal{O}} = 1 - \frac{\mathrm{Li}_\beta(y_c e^{-1/\kappa_{out}})}{\mathrm{Li}_\beta(e^{-1/\kappa_{out}})}; \quad y_c = \phi_1^{OUT}(y_c). \tag{5.40}$$

**Numerical Results**

In order to numerically solve the above equations, the consistency requirement of Eq. (5.34) fixes the relationship between the values $\alpha$, $\beta$, $\kappa_{in}$, and $\kappa_{out}$. In other words, $\kappa_{out} = \kappa_{out}(\alpha, \beta, \kappa_{in})$ and is computed as the zeros of the function

$$K(\kappa_{out}) := \frac{\mathrm{Li}_{\alpha-1}(e^{-1/\kappa_{in}})}{\mathrm{Li}_\alpha(e^{-1/\kappa_{in}})} - \frac{\mathrm{Li}_{\beta-1}(e^{-1/\kappa_{out}})}{\mathrm{Li}_\beta(e^{-1/\kappa_{out}})}, \tag{5.41}$$

i.e., $K(\hat{\kappa}_{out}) = 0$.

By defining a range of values for $\alpha$, $\beta$, and $\kappa_{in}$ the corresponding values of $\zeta$ and $\hat{\kappa}_{out}$ are derived. This means that the $x$-axis values, namely $z = 2\zeta = \langle k_{tot} \rangle$, are fixed. The numerical computation of $\mathcal{L}$, $\widehat{\mathcal{I}}$, and $\widehat{\mathcal{O}}$ is done similarly as described in Sect. 5.1.2, by finding the zeros of the following functions

$$f_{LCC}(x) := \frac{\mathrm{Li}_{\alpha-1}(xe^{-1/\kappa_{in}})\mathrm{Li}_\beta(xe^{-1/\kappa_{out}}) + \mathrm{Li}_\alpha(xe^{-1/\kappa_{in}})\mathrm{Li}_{\beta-1}(xe^{-1/\kappa_{out}})}{2\zeta x \mathrm{Li}_\alpha(e^{-1/\kappa_{in}})\mathrm{Li}_\beta(e^{-1/\kappa_{out}})} - x, \tag{5.42}$$

$$f_{IN}(x) := \frac{\mathrm{Li}_\alpha(xe^{-1/\kappa_{in}})}{\zeta \mathrm{Li}_\alpha(e^{-1/\kappa_{in}})} \frac{\mathrm{Li}_{\beta-1}(e^{-1/\kappa_{out}})}{\mathrm{Li}_\beta(e^{-1/\kappa_{out}})} - x, \tag{5.43}$$

$$f_{OUT}(x) := \frac{\mathrm{Li}_{\alpha-1}(e^{-1/\kappa_{in}})}{\zeta \mathrm{Li}_\alpha(e^{-1/\kappa_{in}})} \frac{\mathrm{Li}_\beta(ye^{-1/\kappa_{out}})}{\mathrm{Li}_\beta(e^{-1/\kappa_{out}})} - x. \tag{5.44}$$

So, for instance, $f_{LCC}(x_c^{LCC}) = 0$ is a solution for $x_c^{LCC} = \phi_1^{LCC}(x_c^{LCC})$. Finally, inserting $x_c^{LCC}$, $x_c^{IN}$, and $x_c^{OUT}$ into Eqs. (5.36), (5.39), and (5.40) for $\mathcal{L}$, $\widehat{\mathcal{I}}$, and $\widehat{\mathcal{O}}$, respectively, yields the various components size. Interestingly, they are trivial: $\mathcal{L} \equiv \widehat{\mathcal{I}} \equiv \widehat{\mathcal{O}} \equiv \mathcal{S} \equiv 1$ and $\mathcal{I} \equiv \mathcal{O} \equiv \mathcal{T} \equiv 0$. In other words, for this type of network, only one giant SCC emerges, making-up the whole LCC, whereas all the other components are zero. This behavior is also validated by network simulations. As an example, for a scale-free network generated with $\alpha = \beta = -2.5, \kappa_{in} = \kappa_{out} = 10$, yielding $\zeta \approx 1.426$.

## 5.2 A Generic Network-Modeling Framework

The discussion in the last sections shows that random networks indeed can exhibit specific, regular bow-tie topologies. However, the range of variety in their components size is quite restricted. Indeed, the peculiarities of real-world complex networks with bow-tie structures, such as the TNC network, call for general mechanisms which allow the various components sizes to be tweaked arbitrarily. In order to achieve this, we move on to analyze link-formation models of networks. In other words, we move away from networks where the nodes are uncorrelated and introduce mechanisms for node-correlation. But first the existing literature on network-formation models is shortly discussed.

### *5.2.1 An Overview of Existing Network Models*

The literature on network-formation models, yielding scale-free networks (see Appendix B.3), can perhaps best be broken down into two main strands.

#### Preferential-Attachment Models

The study presented in (Barabási and Albert 1999) sparked a huge wave of interest and subsequent research. For the first time it was possible to model one of the most ubiquitous properties of real-world complex networks: their scale-free nature. The proposed model was ingeniously simple and consisted of two main ingredients: growth and preferential attachment. In other words, new nodes preferably attach to existing nodes which already have many neighbors. More formally, at each time step $t$ in the network evolution, a new node $i$ with $m$ links is attached to existing nodes $j$ with the probability $\mathcal{P}(i, j) \propto k_j$, with $k_j$ being the degree of $j$. Notice that this preferential-attachment model was originally devised as an undirected network model.

A multitude of generalizations and extensions to this model have been proposed. To name a few:

- adding an initial attractiveness $A > 0$ to nodes, i.e., $\mathcal{P}(i, j) \propto k_j + A$, meaning that also isolated nodes (with $k_j = 0$) can get attached to (Dorogovtsev et al. 2000;
- letting old links disappear (link decay models) (Dorogovtsev and Mendes 2000) or allowing the nodes to age (Dorogovtsev and Mendes 2000);

- adding either a new node or a new link between existing nodes at each step in the network growth (Krapivsky et al. 2001);
- a preferential-attachment mechanism with a generic non-linear attachment probability (Krapivsky and Redner 2001).

We collectively refer to this whole class of models as generalized preferential-attachment models (GPAM).

#### Fitness Model

On the other hand, so-called *fitness models* (Caldarelli et al. 2002; Servedio et al. 2004) build on the idea that the nodes themselves can have an intrinsic degree of freedom, called fitness. This is in strong analogy to the level-3 idea of assigning non-topological state variables to nodes, see Sect. 1.1.1. In detail, each node $i$ has a fitness value $x_i$. The link-formation probability is described by an analytical function $f$ of the fitness values, i.e., $\mathcal{P}(i,j) = f(x_i, x_j)$, where the link is formed if $f \geq \vartheta$, for some threshold $\vartheta$. In essence, the probability distribution density of fitness $\rho(x_i)$ and the functional form of $f$ fully determine the model. The result of these mechanisms is also a scale-free network.

In (Servedio et al. 2004) analytical methods are introduced, allowing the linking function $f$ to be derived from the choice of the probability distribution $\rho$, and vice-versa. As an example, assuming $\rho(x)$ to be log-normal results in the following link-formation probability, assuming $f(x,y) = g(x)g(y)$:

$$g(x) = \sqrt{\frac{\alpha+2}{\alpha+1}\frac{\gamma^{\alpha+1}-\beta^{\alpha+1}}{\gamma^{\alpha+2}-\beta^{\alpha+2}}}\left(\beta^{\alpha+1} + (\gamma^{\alpha+1}-\beta^{\alpha+1})\mathcal{R}(x)\right)^{\frac{1}{\alpha+1}}, \tag{5.45}$$

where $\mathcal{R}(x) = \int_0^x \rho(z)dz$. The parameter $\alpha$ gives the exponent of the scaling-law degree distribution $\mathcal{P}(k) \propto k^{-\alpha}$. The degree is also a function of the fitness distribution and the linking function: $k(x) = N\int_0^\infty f(x,y)\rho(z)dz$. The constants $\beta$ and $\gamma$ are fixed by normalization conditions ($\int_0^\infty \rho(z)dz = 1, g(\infty) = 1$) and are given by $\lim_{x\to 0} k(x) = \beta N$ and $\lim_{x\to\infty} k(x) = \gamma N$. $N$ is the number of nodes in the network.

Originally, the fitness model was devised as an undirected and static model. However, it can be extended to directed networks and incorporate network growth.

#### Various Other Models

As can be expected, the conceivable possibilities of network-evolution models is vast. A short selection is:

- allowing nodes to belong to different types[4] (Söderberg 2002);

[4] This can be seen as a simple implementation of the idea of node fitness.

- models with specific node correlation (Vázquez et al. 2003; Boguñá and Pastor-Satorras 2003);
- weight-driven network dynamics (Barrat et al. 2004a,b);
- coupling the topology with a dynamical processes in the network (Garlaschelli et al. 2007).

### 5.2.2 The Bow-Tie Modeling Framework

From the previous sections it has become apparent that there is a huge variety in the proposed models of network formation. The only unifying theme to be discerned is that the link-attachment mechanism is very often driven by the degree of the nodes. In particular, models aiming at reproducing empirical networks are mostly highly tailored to the specificities and particularities of the considered real-world network. For instance, the many models of the Internet's topology show a very high degree of specialization (Zhou and Mondragón 2004; Siganos et al. 2006; Carmi et al. 2007). This is perhaps not very surprising and indicative of the fact, that there are no universally valid network models. In other words, the nitty-gritty details of each model play a crucial role in the nature of the resulting network being formed.

In focussing on the task at hand, namely devising a flexible network model that is able to reproduce varying bow-tie components size, it is an interesting observation that the existing literature on modeling empirical scale-free networks with bow-tie topologies is quite sparse. (Tadić 2001; Giammatteo et al. 2010). As mentioned, both the WWW and Wikipedia are characterized by large SCCs. In the WWW, the SCC makes up 60–70 % of the nodes, for the English-language Wikipedia, it is about 80 %.

In this thesis we present an additional strand of real-world complex networks displaying bow-tie topologies, coming from economics (Glattfelder and Battiston 2009; Vitali et al. 2011). These ownership networks are characterized by small SCCs and skewed IN and OUT sizes. In detail, the TNC network described in Chap. 4 has an interesting feature regarding the distribution of the degree in the bow tie: the nodes in the IN have high $k^{out}$, the nodes in the OUT high $k^{in}$ and the SCC-nodes large $k^{in}$ and $k^{out}$. This feature hints at the fact that a pure degree-driven network model is probably too rudimentary to reproduce the wealth of observed patterns.

Inspired by the approach taken in (Giammatteo et al. 2010), we propose a generic modeling framework, combining the simple elements of

1. node addition, i.e., network growth;
2. link formation between existing nodes;
3. fitness-dependent preferential attachment.

This setup can be understood as the melding of link/node growth models with fitness models, mentioned in Sect. 5.2.1. By taking a centrality measure as the nodes fitness means that now the topology of the network is driving the distribution of

**Algorithm 2** ModelingFramework($P_n, P_d, m, \mathcal{F}, N$)

**Require:** Initialize seed network
ComputeFitness($\mathcal{F}$)
**while** $number\_of\_nodes < N$ **do**
  BuildNetwork($P_n, P_d, m$)
  ComputeFitness($\mathcal{F}$)
**end while**

fitness, which, in turn, shapes the topology in the next step of the network formation. In essence, the topology and fitness distribution co-evolve.

### The Model Recipe

Algorithm 2 describes the top-level routine of the ModelingFramework(), depending on the probability of adding nodes $P_n$, the link-direction probability $P_d$, the number of links each new node brings to the network $m$, the final network size $N$ and the chosen measure of fitness $\mathcal{F}$, discussed below.

At each step, the subroutine ComputeFitness($\mathcal{F}$) calculates the values of all the nodes associated with the chose fitness. The core subroutine BuildNetwork($P_n, P_d, m$) contains all the modeling intelligence, given in Algorithm 3. In steps 3–16 the new node is attached to the existing network. Steps 18–21 describe the addition of a new link between existing nodes. The subroutines AddNewNode() grows the network by one node and LinkNodes($i, j$) adds a link from nodes $i$ to $j$. Step 19 removes the source node's neighbors from the list of possible destination nodes, as they are already linked to it and hence shouldn't be considered for adding a new link.

The subroutine SelectNodeAccordingToFitness($list\_of\_nodes$) draws a node from $list\_of\_nodes$ according to the distribution of fitness. Following (Ross 2006), this is best simulated on a computer with the pseudo-code implementation:

```
ctot = sum(ag.c);
J=1;
F = ag.c(J);
e = rand * ctot;
while(F<e)
     J = J+1;
     F = F + ag.c(J);
end
```

Where `ag` is the data structure representing the nodes (agents) properties, such as their centrality given by the vector `ag.c`, `rand` draws a uniformly distributed pseudo-random number and `sum(v)` returns the sum of the components of the vector `v`.

**Algorithm 3** BuildNetwork($P_n$, $P_d$, $m$)

```
1: p_n ← DrawRandomVariable()
2: if p_n ≤ P_n then
3:     new ← AddNewNode()
4:     p_d ← DrawRandomVariable()
5:     list_of_nodes ← all_nodes_in_network
6:     number_of_new_links ← 0
7:     while number_of_new_links < m do
8:         old ← SelectNodeAccordingToFitness(list_of_nodes)
9:         if p_d < P_d then
10:            LinkNodes(new, old)
11:        else
12:            LinkNodes(old, new)
13:        end if
14:        list_of_nodes ← list_of_nodes \ {old}
15:        number_of_new_links ← number_of_new_links + 1
16:    end while
17: else
18:    from ← SelectNodeAccordingToFitness(list_of_nodes)
19:    ignore ← {from} ∪ GetNeighborNodes(from)
20:    to ← SelectNodeAccordingToFitness(list_of_nodes \ {ignore})
21:    LinkNodes(from, to)
22: end if
```

This is a very general set of rules to concoct networks. The crucial ingredient is the idea, that the fitness $\mathcal{F}$ of the nodes can be any preferred measure, for instance

- $k^{in}$ or $k^{out}$ (for $k^{in}$ this resembles the model in Krapivsky et al. 2001);
- the Pagerank centrality (described in Appendix B.7), yielding a simpler version[5] of the model in (Giammatteo et al. 2010);
- any of the new measures of centrality introduced in Chap. 2: $\tilde{\nu}^{\text{int}}$ and $\nu^{\text{net}}$, or $\tilde{\nu}^{\text{net}}$ and $\tilde{\nu}^{\text{net}}$ (see Table 2.2).

It should be highlighted that this Modeling Framework thus reflects a true 3-level network approach.

The choices we made in defining the network evolution algorithm are kept as minimal and generic as possible. Future work could aim at tracking the impact of variations in the rules of Algorithm 21. For instance, the link addition description of steps 18–21 could be varied in many ways. The direction could also be probabilistic, introducing another modeling parameter. Or the chosen centrality could be varied for different parts of the algorithm. For instance, selecting nodes according to their in-degree if they are to be the destination of new links, or to their out-degree if they are source nodes.

[5] There the authors have further mechanisms in the link-formation process: with a certain probability, a new link between existing nodes is complemented by an opposing link, next to having a uniform distribution for the number of new links $m$ attached to the new nodes being added.

### 5.2.3 Applying the Framework

To summarize, the Modeling Framework we propose comprises two dynamic elements: new nodes or new links are added in the network evolution. The recipe for choosing existing nodes to attach to is given by the node's importance, reflected in the chosen fitness measure $\mathcal{F}$.

The model is determined by four parameters:

1. the resulting size of the network $N$;
2. the number of links attached to the new nodes $m$;
3. the node/link addition probability $P_n$;
4. the link-direction probability $P_d$.

An additional choice is the shape of the initial seeding network.

#### $\mathcal{F}$ as Degree

The crucial parameters having the most impact on the resulting evolution of the network are $P_n$ and $P_d$. To understand their meaning, we analyze a simple directed variant of the original preferential-attachment model, where the centrality is given by the total degree of the nodes plus a possible constant. In other words, the preferential-attachment probability to chose a node $i$ is

$$\mathcal{P}(i) \propto k_i^{tot} + A = k_i^{in} + k_i^{out} + A. \tag{5.46}$$

Note that the value of $A$ impacts the slope of the emerging scaling-law distribution of the degrees (Dorogovtsev et al. 2000; Dorogovtsev and Mendes 2003).

The simple choice $(P_n, P_d) = (1, 1)$ yields a network evolution which is only driven by node-growth, where all the new nodes point to the existing nodes in the network. Note that this results in $k^{out} \equiv m$. The behavior is equivalent for $(P_n, P_d) = (1, 0)$, switching the direction ($k^{in} \equiv m$). Figure 5.4 shows the degree distributions resulting from different values of $P_d$ for this kind of GPAM, employing Eq. (5.46) with $A = 0$ and $P_n = 1$. When setting $P_d$ to zero or unity, the emerging scaling-law degree probability density is, as expected from the undirected case, described by an exponent $\alpha' = \alpha - 1 \approx 3$. In essence, this corresponds to the simple generalization to directed networks of the original model proposed in (Barabási and Albert 1999) which did not consider bow-tie structures.

Observe, that as long as $P_n$ is fixed to unity, no bow-tie structures can emerge in the whole range of $P_d \in [0, 1]$. This is by construction, then as long as the initial network contains no SCC, no loops can form in the network evolution if every new node has the direction of the $m$ links all aligned in parallel. In contrast, if every new node would be allowed to have the direction of the $m$ links be probabilistic, i.e., if step 4 in Algorithm 3 is moved into the while loop, we then do see the emergence of

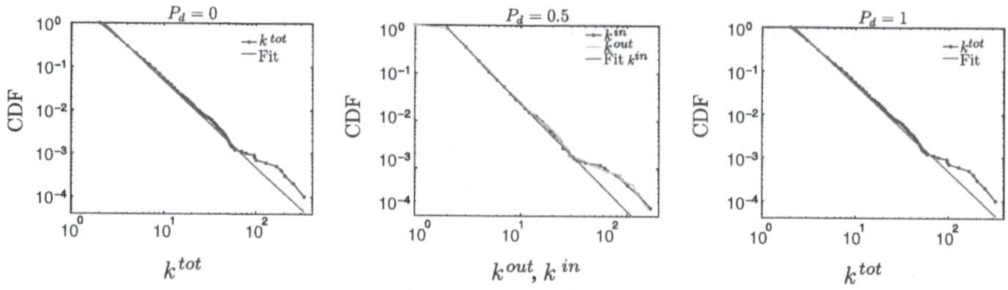


**Fig. 5.4** Degree distributions of the modeling framework with the general preferential attachment probability given in Eq. (5.46), i.e., the GPAM, for $P_n = 1$, $A = 0$, $N = 10000$, $m = 2$ and by varying $P_d$: (*left*) $P_d = 0$, $k^{tot}$ is depicted (note, $k^{in} \equiv m$) and a scaling law with exponent $\alpha = -2$; (*middle*) $P_d = 0.5$, $k^{in}$ and $k^{out}$ are show and a scaling law with exponent $\alpha = -2.25$; (*right*) $P_d = 1$, $k^{tot}$ is shown (note, $k^{out} \equiv m$) and a scaling law with exponent $\alpha = -2$

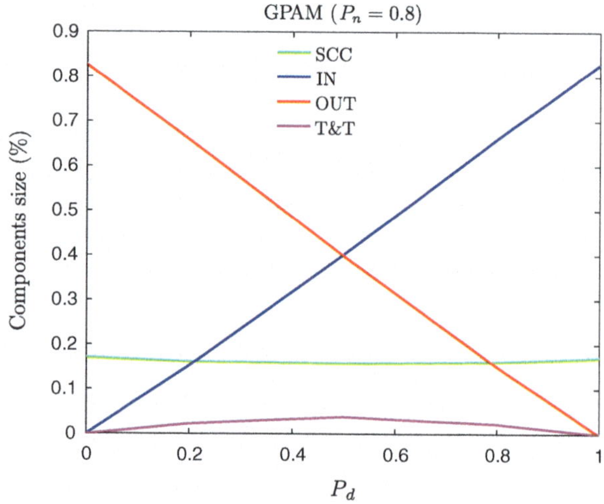


**Fig. 5.5** Bow-tie components size as functions of the link direction $P_d$ for the Modeling Framework with $P_n = 0.8$, $A = 0$ and $m = 2$ using the GPAM described in Fig. 5.4; the values are averages over ten network realizations each containing 10000 nodes

bow-tie topologies for $P_n = 1$. However, as the resulting SCC sizes are very large we don't want to investigate this variant further.

In the proposed Modeling Framework we see the first emergence of a bow-tie topology if $P_n < 1$. This is due to the fact that now new links are added to existing high-fitness nodes, which increases the probability of the emergence of an SCC. Note that the choice $P_n = 0$ is meaningless, as the network size is fixed to the size of the initial network and there is no growth. Figure 5.5 shows the quantitative dependence of the bow-tie components size as a function of $P_d$ for $P_n = 0.8$ for the GPAM employing Eq. (5.46) and $A = 0$. A few things are interesting. The SCC has an approximately constant size, and the IN and OUT sizes have a close-to-linear behavior. There are also no sizable T&T. By setting $P_n$ to a lower value, the SCC size increases at the cost of the other components size.

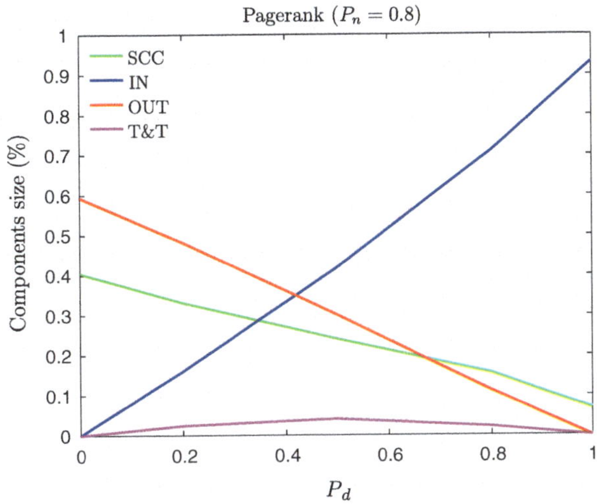


**Fig. 5.6** Bow-tie components size as functions of the link direction $P_d$ for the Modeling Framework with $P_n = 0.8$, $m = 2$ and Pagerank as generalized fitness given in Eq. (5.47); the values are averages over ten network realizations each containing 10000 nodes

These results immediately suggest that this incarnation of the Modeling Framework is already a valid candidate the reproduce the simple bow-tie sizes of the WWW and Wikipedia. The approximate size ranges of the SCC, IN and OUT are, respectively, 60–70 %, 15–20 % and 15–20 % for the WWW (Broder et al. 2000; Donato et al. 2008) and 67–90 %, 5–12 % and 4–16 % for Wikipedia (Capocci et al. 2006). Especially as the parameter $A$ in Eq. (5.46) allows the fine-tuning of the scaling-law exponent of the degree distributions. It is encouraging that our Modeling Framework is indeed generic enough to be applicable to the modeling of two different real-world complex networks.

### $\mathcal{F}$ as Pagerank

To further elucidate the Modeling Framework, we will analyze the behavior given by choosing Pagerank as generalized fitness, detailed in Appendix B.7,

$$\mathcal{P}(i) = pr_i = \alpha \sum_{j \in \Gamma(i)} \frac{pr_j}{k_j^{out}} + \frac{1-\alpha}{N}, \tag{5.47}$$

where $\Gamma(i)$ is the set of labels of the neighboring nodes of $i$. Again, a bow-tie emerges in the network formation for $P_n < 1$. Now the functional $P_d$-dependence of the components size is more complex, as seen in Fig. 5.6. The size of the SCC is now also variable and the asymmetry of the IN and OUT sizes can be explained as follows. For $P_d = 0$ every new node is pointed to by the existing nodes. This results in $k^{in} \equiv m$. By inspecting Eq. (5.47) it is clear that the Pagerank value of a node $i$ depends on the number of neighboring nodes pointing to it. This means that it is very sensitive to the distribution of $k^{in}$, which, if it has a constant value for all nodes, results in a close-to-uniform preferential-attachment mechanism because all nodes have similar Pagerank values. As a consequence, the network is not scale-free anymore and the distribution of $k^{out}$ can be approximated by an exponentially decaying function. In summary, this model's domain of validity is restricted to values of $P_d$ larger than zero and the impact of varying the parameter $P_d$ does not result in a symmetrical behavior as in the degree-dependent case seen in Fig. 5.5. This also means that this incarnation of the Modeling Framework cannot yield small IN sizes. In particular, small SCCs are associated with small OUT and large IN sizes. Similarly to the degree-dependent case, setting $P_n$ to lower values increases the SCC size at the cost of the other components size.

### $\mathcal{F}$ as Network Control or Integrated Control

As a last fitness, we test our Modeling Framework employing the centrality measures inspired by measuring the flow of control in ownership networks (see Chap. 2.9). We will use the network value as preferential attachment probability

$$\mathcal{P} = v^{\text{net}} = (I - W)^{-1} v, \tag{5.48}$$

as seen in Eq. (2.46). $v^{\text{net}}$ can also be interpreted as network control using a linear model to estimate control (see Sect. 2.7. In addition, we also employ the integrated value as probability

$$\mathcal{P} = \tilde{v}^{\text{int}} = (I - W)^{-1} W v, \tag{5.49}$$

given in Eq. (2.42). Again, this measure can also be interpreted as integrated control using the linear model for direct control. Note that if the computation of $(I - W)^{-1}$ happens to be singular in the network evolution (see a discussed in Sect. 2.2.2) we multiply $W$ with a dampening term similar to the one used in Pagerank in Eq. (5.47), with a value of $\alpha = 0.95$: $v^{\text{net}} = \alpha W v^{\text{net}} + v$, and $\tilde{v}^{\text{int}} = \alpha W(\tilde{v}^{\text{int}} + v)$.

As we are interested in the basic behavior of the resulting network evolution, we do not consider the corresponding corrected values $\tilde{v}^{\text{net}}$, defined in Eq. (2.77) and $\tilde{v}^{\text{int}}$, seen in Eq. (2.81). This is kept for future work. Especially as the computational requirements for these corrected centralities are very demanding. However, a preliminary analysis of their behavior revealed that the results are comparable to the uncorrected case. It is an interesting fact that the network evolution is not extremely sensitive to the corrections given in Sect. 2.3.4, whereas the computation of the centralities is.

Figure 5.7 displays the link-direction dependence of the model. The left diagram shows the bow-tie components size for varying $P_d$ for $v^{\text{net}}$ as fitness. The right plot displays this for $\tilde{v}^{\text{int}}$. The first thing that becomes apparent is that the SCC's dependence on $P_d$ is inverted to the case employing Pagerank, as seen in Fig. 5.6. In contrast, the qualitative behavior of the IN and OUT sizes stays comparable. This observation already allows the conclusion, that this embodiment of the Modeling Framework is suited to yield networks with small IN and small SCC sizes. In effect,

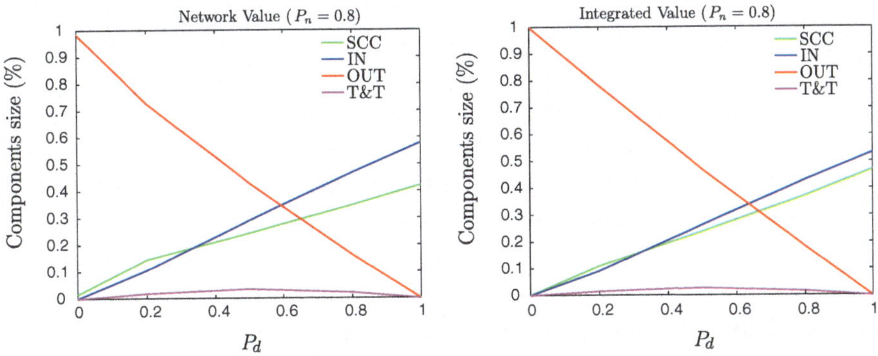


**Fig. 5.7** Bow-tie components size as functions of the link direction $P_d$ for the Modeling Framework with $P_n = 0.8$ and $m = 2$; (*left*) for the network value centrality $v^{\text{net}}$ as fitness; (*right*) for the integrated value centrality $\tilde{v}^{\text{int}}$ as fitness; the values are averages over ten network realizations each containing 10000 nodes

network control or integrated control, interpreted as novel centrality measures, are sensitive to $k^{out}$ due to the fact that the flow of control is against the direction of the links. Very encouraging is that the resulting parameter space of this model encompasses the distinct bow-tie signature observed in ownership networks. Again, as in the two alternative cases discussed above, lowering the value of $P_n$ increases the overall size of the SCC at the cost of the IN and OUT.

It should be pointed out that the Modeling Framework using $\tilde{v}^{\text{int}}$ as fitness has some peculiarities. These are due to the fact that nodes $i$ can have $\tilde{v}_i^{\text{int}} = 0$ resulting in a zero probability of ever being chosen in the preferential-attachment mechanism during the whole formation process of the network. As an example, for $(P_n, P_d) = (1, 0)$, where all new nodes get pointed to by the existing ones, every added node has $\tilde{v}_i^{\text{int}} = 0$ because there is no inflow of control ($k_i^{out} = 0$). This is a very pathological case. On the other hand, for $(P_n, P_d) = (1, 1)$ all new nodes have $k^{out} \equiv m$, resulting in a close-to-uniform distribution of control. Similarly to the Pagerank case for $(P_n, P_d) = (1, 0)$, the resulting network is not scale-free and $k^{out}$ can be approximated by an exponentially decaying function. This reduces the validity of the model to a restricted range of $P_d$ between zero and one, centered around the value of one half.

A final observation is that because $v^{\text{net}} = \tilde{v}^{\text{int}} + v$, this can be interpreted as in the case of the degree-dependent model seen in Eq. (5.46): the preferential-attachment probability $\mathcal{P}(i)$ is augmented by a value $v_i$.

### Overview

To summarize, the parameter $P_n$ is instrumental for the emergence of the bow-tie structure, as it is responsible for the degree the existing high-fitness nodes become connected among each other. Tuning its value down from unity increases the resulting

SCC's size. By tweaking $P_d$, one can determine the size of the IN and OUT sections. For instance, a value close to one produces a huge IN and a small OUT. This is because each new node in the network evolution then points to the existing nodes, which are high-fitness nodes, i.e., either SCC-nodes (their number depending on $P_n$) or IN-nodes, which increases the IN size. Only very few OUT-nodes exist, which formed when the SCC emerged early in the formation process. The T&T appear with a low probability when the new nodes attach to OUT-nodes.

## 5.3 The Bow-Tie Model of Ownership Networks

Now that we have come to terms with the intricacies of the Modeling Framework, the final task is to reproduce some of the empirical properties of the TNC network in the network evolution. This requires two preliminary selections:

1. What empirical features of the TNC network should be captured?
2. Which fitness-dependence should be plugged into the Modeling Framework?

In a final step, embedding the ownership network in the real-world context it pertains to, requires the modeling parameters and mechanisms to be interpreted from an economics point-of-view. This represents the first attempt in the empirical validation of the bow-tie model of ownership networks.

### *5.3.1 Modeling Preliminaries*

#### Empirical Properties

Of the multitude of empirical properties discussed in Sect. 4.2, we focus on the two succinct characteristics of the TNC network:

1. its scale-free nature;
2. the existence of a highly concentrated bow-tie structure with the tiny SCC being comprised of nodes with high network control.

The empirical signature uncovered in Chap. 4 is summarized in Table 5.2. Recall from Sect. 4.2.2 that the data on $k^{in}$ is assumed to be impacted by a systematic bias which is reflected in its peculiar distribution, seen in Fig. 4.3a. For this reason we do not attempt to reproduce the distribution of $k^{in}$.

It is also worth mentioning that the concept of *stylized facts* is not really applicable in our case. As we only have a single-snapshot realization of the TNC network we cannot be too confident in claiming that the measured numerical values are robust enough to represent universal empirical features, i.e., stylized facts. We are, however, confident that qualitatively the tiny SCC comprised of important economic

**Table 5.2** Empirical signature of the LCC in the TNC; for the bow-tie components the values represent percentages with respect to the LCC size (463006 nodes); $\alpha_{out}$ refers to the estimated scaling-law exponent of $k^{out}$; in the last column the average total degree is shown

| IN | SCC | OUT | T&T | $\alpha_{out}$ | $\langle k^{tot} \rangle$ |
|---|---|---|---|---|---|
| 1.213 | 0.285 | 70.099 | 28.403 | −2.15 | 3.358 |

actors, the small IN and huge OUT reflect a very important feature of the TNC network.

#### Choice of Fitness

Concerning the choice of generalized fitness, we will focus on network control (corresponding to network value $v^{\text{net}}$ using a linear model of control). We do not consider $\tilde{v}^{\text{int}}$ as it has undesirable pathologies, as discussed in Sect. 5.2.3. The GPAMs are also not considered, as we do not want a degree-driven network evolution (see more details in Sect. 5.3.4 below). Finally, as also discussed in Sect. 5.2.3, Pagerank can be understood as a diametrically opposing centrality measure to network control.

### 5.3.2 Scanning the Parameter-Space

From the left-hand "phase diagram" seen in Fig. 5.7 it is apparent that the range of interesting parameters are given for $P_n$ close to one and $P_d$ close to zero. Table 5.3 shows a scan of the relevant parameter-space. Note that an SCC size of 0.285 % corresponds to only 28 nodes in a 10000-node network. This represents quite a challenge in a network evolution model. As we are not confident how universal this value is, we are content to reproduce small SCC sizes around 1 % in the final model.

### 5.3.3 The Bow-Tie Model of the TNC Network

We are able to qualitatively mimic the TNC network's empirical signature with the Modeling Framework employing network control as fitness for the parameter values $P_n = 0.99$, $P_d = 0.1$ and $m = 2$. In Table 5.4 the results are summarized and compared with the empirical properties. Figure 5.8 shows the resulting degree distribution of $k^{out}$ which can be fitted by a scaling law with an exponent $\alpha \approx -2$. In Fig. 5.9 four snapshots of the network formation are shown. The initial seeding network is comprised of nodes 1 and 2 with the link $1 \rightarrow 2$. The network at time

**Table 5.3** Averaged statistics over ten network realizations, each resulting network having $N = 10000$ nodes

| $P_n$ | $P_d$ | $\langle k^{tot} \rangle$ | $\alpha_{out}$ | # Links | % LSCC | % IN | % OUT | % T&T |
|---|---|---|---|---|---|---|---|---|
| 0.980 | 0.06 | 4.040 | 1.924 | 20199.4 | 2.318 | 4.123 | 87.313 | 6.246 |
| | 0.07 | 4.040 | 1.922 | 20199.4 | 2.247 | 4.970 | 85.292 | 7.491 |
| | 0.08 | 4.040 | 1.970 | 20199.4 | 2.647 | 5.662 | 84.252 | 7.439 |
| | 0.09 | 4.040 | 2.000 | 20199.4 | 2.561 | 6.408 | 82.320 | 8.711 |
| | 0.10 | 4.040 | 2.265 | 20199.4 | 2.766 | 7.333 | 79.988 | 9.913 |
| | 0.11 | 4.040 | 2.438 | 20199.4 | 2.798 | 8.144 | 78.863 | 10.195 |
| 0.985 | 0.06 | 4.030 | 1.930 | 20150.4 | 2.005 | 4.244 | 86.647 | 7.104 |
| | 0.07 | 4.030 | 1.930 | 20150.4 | 1.945 | 5.104 | 84.230 | 8.721 |
| | 0.08 | 4.030 | 1.926 | 20150.4 | 2.228 | 5.802 | 82.743 | 9.227 |
| | 0.09 | 4.030 | 2.083 | 20150.4 | 2.199 | 6.602 | 80.652 | 10.547 |
| | 0.10 | 4.030 | 2.119 | 20150.4 | 2.281 | 7.430 | 79.302 | 10.987 |
| | 0.11 | 4.030 | 2.381 | 20150.4 | 2.209 | 8.424 | 76.637 | 12.730 |
| 0.990 | 0.06 | 4.020 | 1.938 | 20099.4 | 1.503 | 4.433 | 83.818 | 10.246 |
| | 0.07 | 4.020 | 2.005 | 20099.4 | 1.540 | 5.243 | 81.999 | 11.218 |
| | 0.08 | 4.020 | 1.928 | 20099.4 | 1.579 | 6.069 | 79.950 | 12.402 |
| | 0.09 | 4.020 | 1.928 | 20099.4 | 1.694 | 6.873 | 78.562 | 12.871 |
| | 0.10 | 4.020 | 2.283 | 20099.4 | 1.635 | 7.843 | 75.205 | 15.317 |
| | 0.11 | 4.020 | 2.234 | 20099.4 | 1.738 | 8.721 | 73.655 | 15.886 |

**Table 5.4** Empirical signature of the LCC seen in Table 5.2 and the best-fit achieved with the bow-tie model using the parameters $N = 25000$, $m = 2$, $P_n = 0.99$ and $P_d = 0.1$; the resulting network has 50265 links

| | IN | SCC | OUT | T&T | $\alpha_{out}$ | $\langle k^{tot} \rangle$ |
|---|---|---|---|---|---|---|
| TNC | 1.213 | 0.285 | 70.099 | 28.403 | −2.15 | 3.358 |
| Bow-tie model | 7.720 | 1.344 | 73.840 | 17.096 | −1.922 | 4.021 |

$t = 4$ is shown in the top panel, where the new node and its links are seen in red. In the left-hand network given in the middle panel we see the emergence of the bow tie. For the first time in this network evolution a link is added between two existing nodes resulting in the formation of an SCC of three nodes, seen on the right-hand diagram in black. Finally, on the bottom panel the network comprised of 40 nodes is shown. It is clear, that for this network the high-control nodes (as indicated by the node sizes) are located in the SCC and the IN section. This is in full correspondence with the TNC network.

Recalling that $v^{\text{net}} = \tilde{v}^{\text{int}} + v$, it is an interesting observation that the choice of distribution of $v$ has a small impact on the model's signature. For instance, choosing a log-normal distribution or a uniform one yields similar results. As a final remark, the choice of the initial seeding network does not change the qualitative outcome of the model.

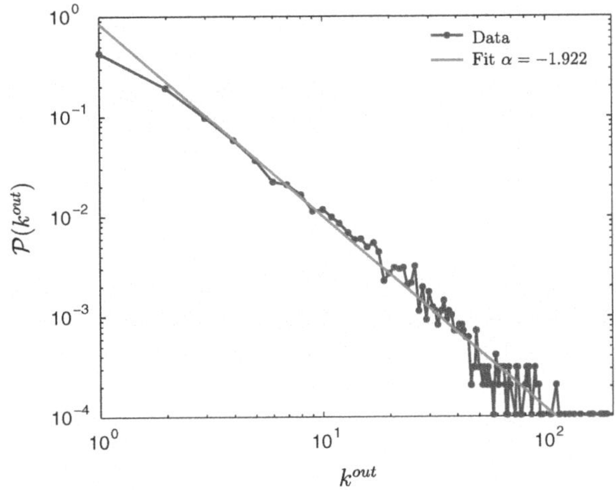


**Fig. 5.8** Degree probability distribution of $k^{out}$ for the bow-tie model, i.e., using the Modeling Framework with $v^{net}$ as centrality, $N = 25000$ and $m = 2$; a scaling law with exponent $\alpha = -1.922$ is shown, fitted with a cutoff at $k^{out} > 55$, $R^2 = 0.9638$

### *5.3.4 Summary and Conclusion: The Economics Interpretation*

The investigation of large-scale economic networks has only slowly started to gain the interest of academics (see also Sect. 1.1.2). The particular challenges of this line of research were recently summarized and discussed (Schweitzer et al. 2009, 2010). In this thesis we uncovered an unsuspected regularity in the organizational structure of the global TNC network: the emergence of a tiny core of interconnected key economic actors, forming an SCC and hence yielding a bow-tie topology. This is in contrast to the few other real-world complex networks exhibiting bow-tie structures where the SCCs are very large (e.g., the WWW and Wikipedia).

From a theoretical modeling perspective, what does this macroscopic feature imply for the microscopic behavior of the actors, reflected in this network structure? In general, most existing models of network evolution are perceived as being too "mechanistic" to be able to capture the true dynamical behavior of nodes and links. Moreover, the development of new dynamic models is understood as a major challenge (Schweitzer et al. 2010).

In static models of network games, a relationship between an economic agent's utility and its network centrality has been shown to exist (Ballester et al. 2006). In detail, the centrality measure of an agent gives it an objective utility, which the agent tries to maximize. In other words, there exists economic situations in which the agents try and maximize their centrality in the network.

Recall that the novel measure of network control is tailored to reflect the importance of economic actors in terms of potential power, see Sect. 6.1.3. It has a precise interpretation as control gained via ownership relations in a directed and weighted ownership network with economic values assigned to the nodes. Crucially, network control is akin to variants of eigenvector centrality, see Sect. 2.6. It is therefore an ideal candidate to act as the utility function that the economic actors are trying to optimize. We therefore conjecture that in ownership networks the agents are striving to maximize their level of network control.

### The Model

The network Modeling Framework introduced above is then an ideal toolkit to consider such microscopic dynamics, as it is conceived as a generic node/link formation model with a preferential-attachment mechanism based on some generalized fitness. This also means that the network evolution model we propose considers all three levels of analysis (see Sect. 1.1.1): node heterogeneity next to directed and weighted links.

By plugging in the network control as fitness, we are able to find a close correspondence between the simulated network signature and the TNC network's empirical one, as seen in Table 5.4. The agreement of the modelled structure with the empirical structure occurs in the following range of the parameters characterizing the link-node formation process: $P_n$ close to one, $P_d$ around 0.1 and $m = 2$. Recall that $P_n = 1$ is a pure node-growth model with no new links being formed between existing nodes and that $P_d = 0$ means that all the new nodes being added to the network get pointed to by the existing ones.

In summary, the network evolution is mostly driven by growth. Most new firms entering the market get owned by the most powerful existing companies. Rarely a new firm gets to own shares in one of the existing high-control corporations, however, instantaneously giving it also a high level of control. As a result, the oldest firms tend to be the most important ones. In addition, with a small but not negligible probability, new links are formed among the most powerful existing firms, resulting in the emergence of the tightly-knit core of influential firms.

This leads to some interesting and falsifiable predictions. The premise of such a high level of growth is contestable. To what extent it is feasible that new firms enter the market at a high rate and how this is applicable for TNCs should be investigated in future studies. At first blush it appears to be a modeling artefact necessary to keep the SCC small.

However, the idea that the oldest companies are the most important ones is very intuitive, as they have had more time to establish themselves in the market. That most new firms have a propensity to be owned by the existing powerful corporations is also not very controversial. If economic agents are indeed optimizing their network control, an easy way to do this, next to establishing links amongst themselves, is to prey for the new firms entering the market.

As seen in the bottom panel of Fig. 5.9, the resulting distribution of high-centrality nodes in the simulated network corresponds to the qualitative distribution in the TNC

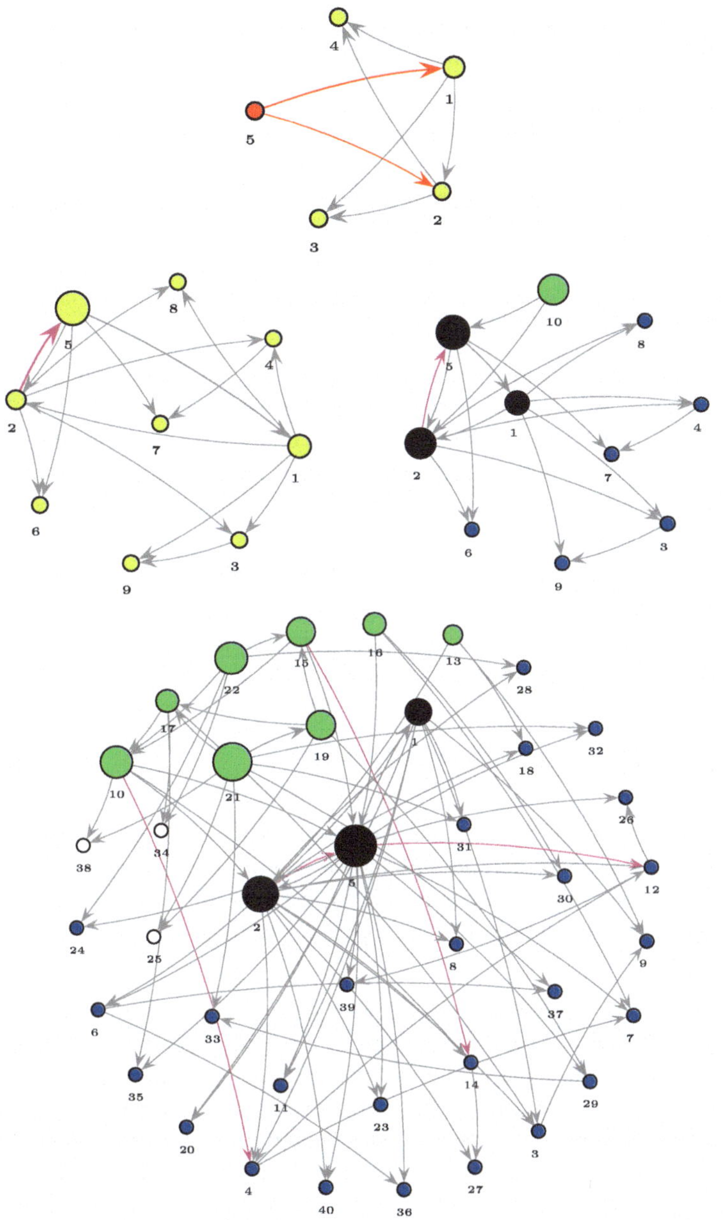

**Fig. 5.9** Various stages in the network evolution; *yellow* nodes denote the network state before the bow-tie emerges; new links between existing nodes are colored *pink*; new nodes are colored *red* with $m$ new *red* links; the bow-tie nodes are represented by *green*, *black*, *blue* and *white* nodes, corresponding to the IN, SCC, OUT and T&T, respectively; the size of the nodes reflects their network value $v^{\text{net}}$; the link thickness represents the weight; the parameter values are $P_n = 0.89$, $P_d = 0.235$ and $m = 2$; see the discussion in the main text (the snapshots were produced using the Cuttlefish Workbench, an open-source project found at http://cuttlefish.sourceforge.net/)

network. Next to the tiny SCC being comprised of the most powerful economic actors, the small IN has, by construction (as network control flows against the direction of links), also high-degree nodes.

In choosing $m = 2$, the number of links associated with every new node, we are able to get the average total degree to correspond to the low value of the TNC network, as seen in Table 5.4. To what degree this strong constraint reflects a true market feature is not clear. In future work it would be advisable to let the number $m$ be probabilistic at every step in the network evolution as well, as suggested in (Giammatteo et al. 2010). Such a level of flexibility is probably more in-line with the actual mechanisms driving the evolution of the TNC network.

Another strong constraint imposed by our model is the fact that the weights of the links have no diversity. At each step in the network-formation process, the distribution of weights is simply determined by normalization, i.e., $\sum_i W_{ij} = 1$, recalling Eq. (1.1). Hence, every node $i$ has incoming links of equal weight. This is in contrast to the empirical distribution of the link weights. This constraint could be eased in future versions of the model. However, the complexity of having varying weights included in the dynamics is a non-trivial task (Barrat et al. 2004a,b).

Finally, it should be investigated why the corrected control centralities do not yield networks with a better correspondence to the TNC network and why the distribution of $v$ appears to play a marginal role in the network evolution.

## In Closing

These results shed new light on the formation of networks of economic agents. They show that such networks are highly non-random and that economic forces are heavily shaping their evolution. In particular, the observed structures are explained by a strong reinforcement mechanism that gives priority of action to the powerful actors.

Because our simple model has very few underlying assumptions (only link and network growth next to a centrality-driven preferential attachment mechanism), this endeavor can be seen as the beginning of an extensive modeling effort of ownership networks. There are many conceivable extensions, refinements and additions possible. However, as we are able to reproduce the key bow-tie signature next to the empirical out-degree scaling-law exponent of the TNC network only with such a simple model and network control as centrality, which is in-line with the economic intuition (Ballester et al. 2006), we are confident to have uncovered a genuine micro-foundation for the empirical pattern that is not an arbitrary modeling artefact.

In essence, the model we present here represents a minimal setup able to capture the TNC network's basic characteristics. It is highly motivating for future work, that this is even possible with such a crude model. Although the various refinements proposed here probably would result in a better correspondence and fewer modeling artefacts, the main result we show here are the following:

> There exists a generic framework for simulating the network evolution which can reproduce the various bow-tie structures observed in real-world complex networks.
>
> Applied to the TNC network, it is possible to empirically validate the model to an encouraging degree.

## References

C. Ballester, A. Calvo-Armengol, Y. Zenou, Who's who in networks. Wanted: the key player. Econometrica, **74**(5):1403–1417 (2006)

A.L. Barabási, R. Albert, Emergence of scaling in random networks. Science **286**, 509 (1999)

A. Barrat, M. Barthélemy, A. Vespignani, Weighted evolving networks: coupling topology and weight dynamics. Phys. Rev. Lett. **92**(22), 228701 (2004a)

A. Barrat, M. Barthélemy, A. Vespignani, Modeling the evolution of weighted networks. Phys. Rev. E **70**(6), 66149 (2004b)

M. Boguñá, R. Pastor-Satorras, Class of correlated random networks with hidden variables. Phys. Rev. E **68**(3), 36112 (2003)

A. Broder, R. Kumar, F. Maghoul, P. Raghavan, S. Rajagopalan, S. Stata, A. Tomkins, J. Wiener, Graph structure in the web. Comput. Netw. **33**, 309 (2000)

G. Caldarelli, A. Capocci, P. de Los Rios, M.A. Muñoz, Scale-free networks from varying vertex intrinsic fitness. Phys. Rev. Lett. **89**(22), 258702 (2002)

A. Capocci, V.D.P. Servedio, F. Colaiori, L.S. Buriol, D. Donato, S. Leonardi, G. Caldarelli, Preferential attachment in the growth of social networks: the internet encyclopedia wikipedia. Phys. Rev. E **74**(3), 36116 (2006)

S. Carmi, S. Havlin, S. Kirkpatrick, Y. Shavitt, E. Shir, A model of internet topology using k-shell decomposition. Proc. Natl. Acad. Sci. **104**(27), 11150 (2007)

D. Donato, S. Leonardi, S. Millozzi, P. Tsaparas, Mining the inner structure of the web graph. J. Phy. A Math. Theor. **41**(22), 224017 (2008)

S.N. Dorogovtsev, J.F.F. Mendes. Evolution of networks with aging of sites. Phys. Rev. E **62**(2), 1842–1845 (2000a)

S.N. Dorogovtsev, J.F.F. Mendes, Scaling behaviour of developing and decaying networks. EPL **52**, 33–39 (2000b)

S.N. Dorogovtsev, J.F.F. Mendes, *Evolution of Networks: From Biological Nets to the Internet and WWW* (Oxford University Press, Oxford, 2003)

S.N. Dorogovtsev, J.F.F. Mendes, A.N. Samukhin, Structure of growing networks with preferential linking. Phy. Rev. Lett. **85**(21), 4633–4636 (2000)

S.N. Dorogovtsev, J.F.F. Mendes, A.N. Samukhin, Giant strongly connected component of directed networks. Phys. Rev. E **64**(2), 25101 (2001)

R.A. Durrett, *Probability Theory and Examples* (Thomson Learning, London, 2004)

D. Garlaschelli, A. Capocci, G. Caldarelli, Self-organized network evolution coupled to extremal dynamics. Nat. Phys. **3**(11), 813–817 (2007)

P. Giammatteo, D. Donato, V. Zlatić, G. Caldarelli, Pagerank Based Preferential Attachment Model for the Evolution of the WWW (Forthcoming; privately obtained from G, Caldarelli, 2010)

J.B. Glattfelder, S. Battiston, Backbone of complex networks of corporations: the flow of control. Phys. Rev. E **80**(3), 36104 (2009)

P.L. Krapivsky, S. Redner, Organization of growing random networks. Phys. Rev. E **63**(6), 66123 (2001)

P.L. Krapivsky, G.J. Rodgers, S. Redner, Degree distributions of growing networks. Phys. Rev. Lett. **86**(23), 5401–5404 (2001)

M.E.J. Newman, S.H. Strogatz, D.J. Watts, Random graphs with arbitrary degree distributions and their applications. Phys. Rev. E **64**(2), 26118 (2001)

S.M. Ross, *Introduction to Probability Models* (Academic Press, San Diego, 2006)

F. Schweitzer, G. Fagiolo, D. Sornette, F. Vega-Redondo, A. Vespignani, D.R. White, Economic networks: the new challenges. Science **325**(5939), 422 (2009)

F. Schweitzer, G. Fagiolo, D. Sornette, F. Vega-Redondo, D.R. White, Economic Networks: What Do We Know and What Do We Need to Know? (Forthcoming in, Advances in Complex Systems, 2010)

V.D.P. Servedio, P. Buttà, and G. Caldarelli. Vertex intrinsic fitness: how to produce arbitrary scale-free networks. Phys. Rev. E, **70**, 056126 (2004)

G. Siganos, S.L. Tauro, M. Faloutsos, Jellyfish: a conceptual model for the AS onternet topology. J. Commun. Netw. **8**(3), 339 (2006)

B. Söderberg, General formalism for inhomogeneous random graphs. Phys. Rev. E **66**(6), 66121 (2002)

B. Tadić, Dynamics of dDrected graphs: the world-wide web. Physica A **293**(1–2), 273–284 (2001)

A. Vázquez, M. Boguná, Y. Moreno, R. Pastor-Satorras, A. Vespignani, Topology and correlations in structured scale-free networks. Phys. Rev. E **67**(4), 46111 (2003)

S. Vitali, J.B. Glattfelder, S. Battiston, The network of global corporate control. PloS ONE **6**(10), e25995 (2011)

S. Wolfram, *A New Kind of Science* (Wolfram Media, Champaign, 2002)

S. Zhou, R.J. Mondragón, Accurately modeling the internet topology. Phys. Rev. E **70**(6), 66108 (2004)

# Chapter 6
# Conclusions

*I think the next century will be the century of complexity.*

(S. Hawking cited in Chui 2000, p. 29A)

The ensuing global financial, economic and sovereign debt crisis makes one thing quite obvious: the understanding of market mechanisms and of the stability of economies is incomplete. To quote Jean-Claude Trichet, then Governor of the European Central Bank, in November 2010 (Farmer et al. 2012):

> *When the crisis came, the serious limitations of existing economic and financial models immediately became apparent. Macro models failed to predict the crisis and seemed incapable of explaining what was happening to the economy in a convincing manner. As a policy-maker during the crisis, I found the available models of limited help. In fact, I would go further: in the face of the crisis, we felt abandoned by conventional tools.*

Why did no one appear to predict the crisis, or, as the title of an article of P. Krugman in the New York Times put it, "How Did Economists Get It So Wrong" (Krugman 2009)? His short answer is:

> *They [economists] turned a blind eye to the limitations of human rationality that often lead to bubbles and busts; to the problems of institutions that run amok; to the imperfections of markets—especially financial markets—that can cause the economy's operating system to undergo sudden, unpredictable crashes; and to the dangers created when regulators don't believe in regulation.*

The previous few voices challenging the *status quo* in economics, mostly stifled by its seeming success, have now continually gained backing by more and more skeptically-minded people. One main critique has been the economists reliance on axioms and oversimplified models. For instance, classical economics is built on very strong assumptions: the rationality of economic agents (to maximize their profits), the "invisible hand" (that agents, in the pursuit of their own profit, are led to do what is best for society as a whole) and market efficiency (that market prices faithfully reflect all known information about assets). Basic concepts are postulated, such as the market equilibrium, which are only relevant when all or most assumptions hold. The advantage of equilibrium models, which contain no evolutionary or adaptive

J. B. Glattfelder, *Decoding Complexity*, Springer Theses,
DOI: 10.1007/978-3-642-33424-5_6, 

dynamics at all, is that they can be solved analytically. Moreover, many economic models aim at solely optimizing some utility function of some representative agent.

In the words of Lord Turner, head of the U.K. financial services authority (Farmer et al. 2012):

> *But there is also a strong belief, which I share, that bad or rather over-simplistic and overconfident economics helped create the crisis. There was a dominant conventional wisdom that markets were always rational and self-equilibrating, that market completion by itself could ensure economic efficiency and stability, and that financial innovation and increased trading activity were therefore axiomatically beneficial.*

An additional point of criticism concerns many classical economists disregard of empirical data. In other words, the concepts in economics are seen as being so strong that they supersede any empirical observation. This is in stark contrast to a foundational guideline principle in science: "let the data speak!"

Although, slowly, alternative fields in economics are incorporating a more data-driven approach. Behavioral economics (Kahneman 2003) aims at understanding human behavior in terms of psychological and cognitive traits, grappling with the ensuing irrationalities in decision-making (Ariely 2008). Other fields are, for instance, evolutionary and information economics next to economic geography. However, these approaches represent a minority view and their insights do not appear to be incorporated into mainstream economics. Although, Joseph Stiglitz argues in a newspaper article that this is slowly changing (Stiglitz and Akerlof 2009). Time will tell if this optimistic view prevails, as the resistance to new ideas can be considerable as exemplified by the quote of the father of the efficient market hypothesis,[1] Eugene Fama, apparently oblivious to the reality of the current crisis, taken from Fama (2010):

> *I don't know what a credit bubble means. I don't even know what a bubble means. These words have become popular. I don't think they have any meaning.*

Or as Mark Buchanan, physicist and author, summarized in Buchanan (2009):

> *In an essay in The Economist, Robert Lucas, one of the key figures behind the present neo-classical theory of macroeconomic systems, even argued that the tumultuous events of the recent crisis can be taken as further evidence supporting the efficient-markets hypothesis of neo-classical theory, despite the fact that it disputes the possible existence of financial bubbles.*

A quote from Jean-Philippe Bouchaud, physicist and pioneer in econophysics, from Bouchaud (2008), perhaps highlights the status of the study of economics with respect to science:

> *Compared with physics, it seems fair to say that the quantitative success of the economic sciences has been disappointing.*
>
> *To me, the crucial difference between modeling in physics and in economics lies rather in how the fields treat the relative role of concepts, equations and empirical data.*

Again quoting Jean-Claude Trichet in Farmer et al. (2012):

---

[1] See Fama (1970).

> *[...] we need to develop complementary tools to improve the robustness of our overall framework. In this context, I would very much welcome inspiration from other disciplines: physics, engineering, psychology, biology. Bringing experts from these fields together with economists and central bankers is potentially very creative and valuable. Scientists have developed sophisticated tools for analysing complex dynamic systems in a rigorous way. These models have proved helpful in understanding many important but complex phenomena: epidemics, weather patterns, crowd psychology, magnetic fields.*

Finally, people have argued that in economics, the focus still lies too much on aspects of individual things and not on the big picture given by the interconnected and networked nature of our world. Such a change in perspective would constitute a fundamental paradigm shift.[2] Understanding a systems components' individual properties does not bring insights into how the system will behave as a whole. Indeed, the very concept of emergence[3] fundamentally challenges our knowledge of complex systems: self-organization allows for novel properties to emerge, features not previously observed in the system or its components. The whole is literally more than the sum of its parts.

Andy Haldane, executive director of financial stability at the Bank of England in London, recently pointed out:

> *Finance is a classic complex, adaptive system, similar to an ecosystem. The growth in its scale, complexity and adaptation in the past generation alone would rival that of most other complex systems in the past century.*
>
> *Yet this dense cat's cradle of finance has been woven largely out of sight. At best we are able to snatch passing glimpses of it, for data are incomplete, local and lagging. Making sense of the financial system is more an act of archaeology than futurology.*
>
> *How so? Because, at least historically, finance has not thought of itself as a system. Instead, financial theory, regulation and data-collection has tended to focus on individual firms. Joining the dots was in no one's job description.*
>
> *Theorists were partly to blame. Economics has always been desperate to burnish its scientific credentials and this meant grounding it in the decisions of individual people. By itself, that was not the mistake. The mistake came in thinking the behaviour of the system was just an aggregated version of the behaviour of the individual. Almost by definition, complex systems do not behave like this.*
>
> *Interactions between agents are what matters. And the key to that is to explore the underlying architecture of the network, not the behaviour of any one node. To make an analogy, you cannot understand the brain by focusing on a neuron—and then simply multiplying by 100 billion.*

See Sect. F.3.2 in the Appendix for the entire article, as it was prompted by our study (Vitali et al. 2011).

So, what could a paradigm shift towards an understanding of economics as a complex system offer? As mentioned in Sect. 1.1.2, the study of economic networks is only just beginning and is usually not performed by economists, see Sect. 1.2.3. This thesis gives a first glimpse of the possibilities of what knowledge can be obtained

---

[2] See Appendix A.1.3.

[3] See Sect. 1.1.

by analyzing vast amounts of economic data in terms of complex networks. On a grander scale of things, the recent EU, US and UN projects, mentioned at the end of Sect. 1.3, are fully dedicated to this new paradigm of viewing financial markets and the economy from a complex-networks point-of-view, driven by empirical data. This perhaps implies that in the not too distant future the kind of research found in this thesis will be part of an established field in interdisciplinary research, extending the scope of Steven Strogatz's prophecy (Strogatz 2001):

> *In the long run, network thinking will become essential to all branches of science as we struggle to interpret the data pouring in from neurobiology, genomics, ecology, finance and the Word-wide Web.*

On a personal note, I hope for the following. Ideas relating to economics, finance, politics and society are very often tainted by individual ideologies. In contrast, decoding the complexity of our world by considering its interconnected and interactive nature, not only brings novel understanding, but perhaps also allows for a neutral perspective to emerge. Reality is so complex, we need to move away from dogma.

The possible real-world relevance of our findings is discussed in the next section, next to addressing some of the criticism we faced. In particular, the results of our empirical analysis should have important implications for issues relating to market competition (Sect. 6.2.1) and financial systemic risk (Sect. 6.2.2). Section 6.3 touches on issues related to future work. In closing, Sect. 6.4 summarizes the contributions of this thesis, with respect to economics and the study of complex networks.

## 6.1 Real-World Relevance

The interest in the topic of our work is witnessed by the media coverage the publications (Glattfelder and Battiston 2009; Vitali et al. 2011) received, see Appendix F. But do any of our results really have any relevance? In the following, we would like to address some of the voiced concerns and common misconceptions of our work.

### *6.1.1 Control from Ownership*

> *Control cannot really be assessed from ownership.*

Although there are many pitfalls in estimating control from ownership, scholars in the field of corporate governance believe that the inferred control from ownership relations is in fact a good proxy for uncovering the otherwise unobservable true level of control (Brioschi et al. 1989; La Porta et al. 1998, 1999; Claessens and

Djankov 2000; Nenova 2003; Chapelle and Szafarz 2005; Chapelle 2005; Almeida and Wolfenzon 2006; Almeida et al. 2007).

Many conceivable issues can be seen to skew or to muddy the control based on ownership:

- the many ways in which cash-flow rights can map to voting rights: nonvoting shares, dual classes of shares, multiple voting rights, golden shares, voting right ceilings, etc.;
- the existence of proxy votes (see Sect. 1.2.1);
- the possibility of shareholder coalitions forming in the voting process;
- the use of complex ownership patterns as vehicles to separate ownership from control (e.g., cross-shareholdings and pyramids, see Sect. 1.2.2).

However, there are directives in many countries aiming for the one-share one-vote principle. In fact, empirical studies indicate that in many countries the corporations actually don't tend to exploit all the opportunities allowed by national laws to skew voting rights (La Porta et al. 1999). Especially in Europe, on average 69 % of companies follow the one-share one-vote principle (The Deminor Group 2005).

Moreover, we compare three drastically different models for computing control from ownership relations[4] (Sects. 2.7 and 2.8). The study seen in Chap. 4 indicates that the results are invariant with respect to the chosen model for estimating control: the overall concentration of control seen in Fig. 4.12 and the identification of the key economic actors, next to the distribution of control in the network structure, as seen in Tables 4.3 and 4.4, are very similar in all three cases. This is a possible validation of the claim, that it is indeed possible to unequivocally estimate control from ownership.

The methodology described in Chap. 2 is especially attuned to track the flow of control in any conceivable network structure. Finally, proxy voting can be seen as increasing the real concentration of control, as financial institutions get to hold blocks of votes of their many shareholders.

### 6.1.2 Cross-Country and Global Analysis

*Comparing ownership, and hence control, in different countries is like comparing apples with oranges.*

There are many conceivable national determinants affecting ownership relations: legal settings, law enforcement, level of corruption, tax rules, institutional settings, market size, maturity of the banking sector, etc. However empirical studies have suggested that the distinguishing feature is in fact mainly the consequence of legal protection (La Porta et al. 1999), see Sect. 1.2.2. It is a positive sign, that the national

[4] Our new model also considers shareholder coalitions.

influences on ownership networks can be ignored in favor of a single universal determinant.

Moreover, one would expect that each of the three models of estimating control amplify different national characteristics in different countries. As mentioned in the last section, the aggregated results we present do not show this expected variability of the models.

In any case, our studies should be understood as the first attempt in uncovering national and global corporate control.

### 6.1.3 Mutual Funds and Control

> *They've [Battiston and Glattfelder] just discovered that funds managers and market-maker banks own a lot of shares! What a scandal!*[5]

What is indeed obvious and not surprising, is the fact that large mutual funds own many shares. However, what we are saying is that the control (inferred from the direct and indirect ownership) they have is surprisingly big. If you are a portfolio manager, you would arguably go for a diversification strategy and not a controlling one.

Although it is perhaps intuitive to think of mutual funds as also having a high degree of control, this notion is not at all demonstrated in the literature. In fact, there is a debate in the scholarly work on the subject if financial institutions exert any kind of control over the firms in their portfolio or not. Hence the main objection is that we are possibly assigning high control to some financial institutions which may not have an interest in exerting it.

It is known that some of the largest US mutual funds,[6] when operating in the US, do not always seek to exert overt control (Davis and Kim 2007; Davis 2008). Control in this case is understood as the mutual fund's propensity to vote against the management of the owned firms. In other words, control is only understood as the voting related to issues of corporate governance in this context. However, the same mutual funds are also described as exerting their power when operating in Europe (Davis and Kim 2007; Davis 2008). To our knowledge, there are no systematic studies about the control of financial institutions over their owned companies world-wide.

Furthermore, we interpret our quantitative measure of control as a measure of "potential power". This notion is somewhat reminiscent of Weber's definition of power, namely the probability of achieving one's own interest against the opposition of other actors (Weber 1997). In this sense, a shareholder has control associated with the probability of achieving his interest when demanding changes in the firms he owns, without stating how this is exerted. For example, a mutual fund owning a significant percentage of a large corporation may try to impose job cuts because

[5] From the blog post "OBVIOUS tag, where are you?": http://josh.sg/2009/02/obvious_tag_where_are_you_1.html.

[6] Often only Fidelity is analyzed in the literature.

of a weak economic situation. This can happen: (i) without actually voting and (ii) although the fund does not plan to keep these shares for many years. In this case, the influence of the mutual fund has a direct impact on the company and its employees. Moreover, mutual funds with shares in many corporations may try to pursue similar strategies across their entire portfolio. In any case, there are only 49 mutual funds (identified by the NACE code 6713) among the 737 top power-holder.

Finally, in the US, banking is generally assumed to be separate from commerce because, unlike for instance in Germany or Japan, banks are barred from making equity investments in non-financial firms for their own account. However, to the contrary, there is strong evidence that US banks do in fact control important voting stakes of firms as a result of the equity investments they make through their trust business. Moreover, bankers are more likely to join the corporate board of a firm in which their bank holding company controls a large voting stake. See details in Santos and Rumble (2006).

Given these premises, we cannot generally exclude that the top shareholders having vast potential power do not globally exert it in some way. In effect, this means that the decisions of a few individuals (corporate managers) can have a great global influence on hundreds of large firms and hence on a lot of people.

### 6.1.4 The Importance of Natural Persons

*But where are today's Rothschild's and Rockefeller's?*

Sometimes the absence of natural persons in the list of top national and international power-holders (Sect. 3.5.2 and Appendix E) was criticized. It is indeed an interesting observation that the power of individuals and their families has faded in the wake of the ascent of the global corporation.

In Sect. 3.5.2, the natural persons appearing in the national backbones was discussed. In the US, Warren E. Buffet and William Henry Gates III are at rank 9 and 26, respectively. In DE, the family Porsche/Piech and in FR the family Bettencourt are power-holders in the top ten. In Switzerland, André Hoffmann and Adreas Oeri are at rank 4 and Thomas Schmidheiny at rank 14. GB and JP have no natural persons in the top ranks.

In the global TNC network, Charles B. Johnson is the first individual at rank 79. Other names further down the list are: the Ford family, Rupert H. Johnson, Warren E. Buffet, François M. Pinault, André Hoffman and Adreas Oeri, Jean-Charles Naouri, Paul Desmarais, the Family Dreyfus and Harold B. Smith.

### 6.1.5 Concentration of Control

*But the idea that economic control is concentrated is well known.*

The concentration of control has, in fact, never been estimated quantitatively, at any scale, previously in the literature. Our studies offer new insights and show how these very high degrees of concentration are associated with various cohesive groups, the top-shareholders identified in Chaps. 3 and 4. However, what the implications of these findings are is beyond the scope of our work and we hope that other scholars will analyze this in future studies.

### Globally

People and scholars have a qualitative intuition that control should be concentrated because wealth is concentrated. Thus we have compared our measure of concentration of control to measures of concentration of economic wealth, based, e.g., on income or revenue, see Sect. 4.3.3.

To summarize: 80 % of the total income is held by 5–10 % of the economic actors. In contrast, we find that 80 % of the total control is held by only 0.61 % of the shareholders. This is what is very surprising and new. It represents a one-order-of-magnitude higher concentration level. It means that, thanks to their links in the network, the top ranked economic actors achieve control 10 times higher than what could be expected based on their wealth. To our knowledge, such a high level of concentration is unprecedented in economics.

### In the US

Regarding the issue of the global concentration of control observed in the US, despite the presence of widely held firms (see Sect. 3.5.1), we are aware of only one study predicting this outcome: (Davis 2008).

We continue here with the story of the history of ownership and control in the US, which was left off in Sect. 1.2.2. Ironically, just as the theoretical case for dispersed ownership was being made by academics, corporate ownership in the US was becoming more concentrated in the hands of institutional investors. The proportion of shares owned by institutions increased from about 35 % in 1980 to almost three-quarters by 2005.

Over half of the American households participated in the stock market in 2001. They did this not through direct ownership but via shares held in mutual funds. Accompanying this development, there has been a shift of American retirement funds in the 1990s away from defined retirement plans (associated with particular employers) to defined contribution plans (invested in mutual funds, e.g., 401k plans). Hence big institutional block-holders emerged. This is also called the "new finance capitalism" in the US: very few mutual funds are the most significant large-scale corporate owners. For instance, Fidelity is the single largest shareholder of 10 % of corporate America. "This is a concentration of corporate ownership not seen since the early days of [bank-centered] finance capitalism" (Davis 2008).

In essence, our study can bee seen as the systematic, exhaustive and international extension of the work done in Davis (2008), where only Fortune 1000 companies and a hand-full of big mutual funds were analyzed.

### 6.1.6 The Economic "Super-Entity"

*Should the core of the bow-tie really be understood as an economic 'super-entity'?*

Notice that the very existence of such a small, powerful and self-controlled group of financial TNCs (see Sect. 4.3.5) was unsuspected in the economics literature. Indeed, its existence is in stark contrast with many theories in corporate governance (Dore 2002).

It is a novel finding that powerful companies do not conduct their business in isolation but are tied together in an extremely entangled web of control. "It's disconcerting to see how connected things really are," agrees George Sugihara, a complex systems expert who has advised Deutsche Bank (from Sect. F.3.1 in the Appendix). This raises issues regarding the desirability of such a structure for the efficiency of the market and its impact on inequality, see Sect. 6.2.1.

Dan Braha of NECSI suspects that the members of the "super-entity" will compete in the market with each other, but could act together on common interests. Resisting changes to the network structure may be one such common interest (from Sect. F.3.1 in the Appendix).

*But isn't this core just an ad hoc assemblage of firms with no strategic driving force?*

Previous studies have investigated the impact of small ownership structures at national level on market competition. For instance, the JP and the US automobile industries (Alley 1997), the global airline industry (Gilo 2000), the financial sector in NL (Dietzenbacher et al. 2000), the Nordic power market (Amundsen and Bergman 2002), and the global steel industry (Gilo et al. 2006).

Thus, there is evidence that even small national groups of firms tied up by cross-shareholdings can act as a bloc. This is why the national antitrust agencies watch them closely. The prospect of an international bloc of financial corporations raises this issue to a new level, as there are no corresponding global counter-parties monitoring such activities. This opens the door to speculations of what global impact this structure could potentially have.

In a recent development, the study (Battiston et al. 2012b) used previously confidential data on the 2008 crisis gathered by the US Federal Reserve (FED emergency loans program) to assess systemic risk. The analysis introduces a novel centrality measure tracking systemic risk, called DebtRank, inspired by the methodology employed in

Chap. 4 (and Pagerank, see Appendix B.7). The debt data is augmented with the ownership data used in Chap. 4, and includes detailed daily balance sheets for 407 institutions that between them received bailout funds worth USD 1.2 trillion from the Federal Reserve. The data covers 1000 days from before, during and after the peak of the crisis, from August 2007 to June 2010. The study focused on the 22 banks that collectively received three-quarters of that bailout money. Almost all were members of the "super-entity".

### 6.1.7 Are the Results Really New?

> *The results are in general not really surprising and a natural fact. Surely they are not new.*

For instance, according to a common intuition among scholars and in the media, the global economy is dominated by a handful of powerful transnational TNCs. However, this has not been confirmed or rejected with explicit numbers.

The list of top power-holder (Appendix E) does not contain any surprising names. Most of the companies belonging to the top list are among those one could expect to see, based on the size of their business. What is new however, is the quantification of the share of control each key economic actor holds and their level of interconnectedness, i.e., the emergence of the economic "super-entity", discussed above.

There is a big difference in suspecting the existence of a regularity and in empirically demonstrating it. As in all fields of science, classification and quantification are the first necessary steps in the process of in-depth understanding of novel phenomena.

To quote the physicist J. Reichardt[7] (see Sect. F.3.3):

> *It's interesting that you can get these results that, if you asked an experienced economist they'd probably have a gut feeling about, but now you can show it in a quantitative way. They've [(Glattfelder and Battiston, 2009)] done a great job of making it mathematically rigorous.*

In the same vein, economist M. Jackson[8] commented on the results presented in Glattfelder and Battiston (2009) in an interview (see Sect. F.3.4):

> *'It's clear, looking at financial contagion and recent crises, that understanding interrelations between companies and holdings is very important in the future,' he [Jackson] said. 'Certainly people have some understanding of how large some of these financial institutions in the world are, there's some feeling of how intertwined they are, but there's a big difference between having an impression and actually having ... more explicit numbers to put behind it.'*

[7] http://theorie.physik.uni-wuerzburg.de/~reichardt/.

[8] http://www.stanford.edu/~jacksonm/.

### 6.1.8 Self-Organization Versus a Global Elite

> *Your results prove that there is a puppet-master behind the scenes, orchestrating this large concentration of power for self-enrichment.*

On the contrary, although our numbers do show an unprecedented high level of concentration, the simulations of Chap. 5 suggest that this is in fact the result of the interaction rules in the system. Contrary to common intuition, it is not necessary to have a global elite conspiring behind the scenes to rule the world, to end up with such a large concentration of power. Inequality can be an emergent property. So, the "super-entity" may not result from conspiracy.

This implies that the very mechanisms of how participants interact in a market according to local rules results in a high level of global concentration. A recent study suggests that market exchange rules are responsible for wealth concentration (Iglesias and de Almeida 2011).

To quote again from one of the articles discussing our work (Sect. F.3.1 in the Appendix):

> *The Zurich study, says Sugihara, "is strong evidence that simple rules governing TNCs give rise spontaneously to highly connected groups". Or as Braha puts it: "The Occupy Wall Street claim that 1 per cent of people have most of the wealth reflects a logical phase of the self-organising economy."*

However, natural and logical does not necessarily mean desirable…

## 6.2 Summary of Possible Implications

Having established the relevance of our findings, in the following, we would like to summarize the possible implications of our work. The main implications, we believe, come from the identification of the core of the network: the SCC of the TNC network acting as an economic "super-entity". Economic theory doesn't suspect its existence, however predicts that this kind of structure should have profound implications in two domains which are, today, in all newspapers: market competition and systemic risk.

### 6.2.1 Market Competition

Previous studies (on small samples) have shown that cross-shareholdings significantly reduce competition (O'Brien and Salop 1999, 2001; Gilo et al. 2006; Trivieri 2007). Consequently, antitrust institutions all around the world (e.g., the UK Office of Fair Trade and the OECD[9] take the existence of complex cross-shareholding

[9] And its Competition Committee, see http://www.oecd.org.

structures very seriously. However, they lack the analytical and quantitative tools to deal with large networks.

Indeed, this probably explains why we have been contacted by a governmental agency regarding the publication (Glattfelder and Battiston 2009) regarding these issues.

### 6.2.2 Financial Systemic Risk

For years economic theory has supported the idea that more connected networks are more stable. High connectivity enables risk sharing which promotes stability. In contrast, recently, the work of some scholars as well as the view of some authoritative policymakers predicts that higher levels of interconnections among financial institutions can lead to higher systemic risk (Battiston et al. 2007, 2012a; Lorenz and Battiston 2008; Wagner 2009; Haldane 2009; Stiglitz 2010).

In essence, if the global financial sector is all tied up together, then financial distress propagates like a contagious disease to all the major institutions. Simulations suggest that there is a threshold effect associated with the level of connectivity. Above a certain value, feedback loops and amplifications can lead to a knife-edge property, where stability is suddenly overshadowed by a systemic crisis.

This is of course extremely undesirable for the economy of the entire world. But it is also precisely what happened in the recent financial turmoil. It follows that, in the debate over financial stability, our results are of great interest, especially because, surprisingly, there is no prior investigation of how entangled the global financial sector is.

In the wake of the financial crisis the aphorism "too big big to fail" was coined, implying that certain financial institutions have become so large that their failure will be disastrous to the economy, hence necessitating the support by governments. However, this is only the first step when thinking in terms of systemic risk and the propagation of financial distress. It's not just size that matters when it comes to how much risk a troubled bank poses to the financial system. The study described in Chap. 4 implies that the next level of trouble is more subtler, where the problem could be identified as "too connected to fail".

In a recent study (Battiston et al. 2012b), where the details are discussed at the end of Sect. 6.1.6, the full picture is perhaps uncovered, described by the colloquial term "too central to fail". The work devises a metric, called DebtRank, to compute the likelihood that a bank will default, and how much this would damage the creditworthiness of other banks and companies in the network. In short, DebtRank proposes a quantitative method for monitoring institutions in a network and identifying the ones that are the most important to the stability of the system. This, finally, is a first possible tangible result from trying to understand the economy as an interconnected complex system.

## 6.3 Future Work

We believe to have only scratched the surface regarding the analysis of ownership relations and the possibilities of 3-level complex network analysis. In this sense, there is a multitude of possible further work to be tackled if one takes the idea seriously, that also in economics, the objects of study should be understood as complex systems, driven by empirical data.

Generally, our empirical work should be embedded in a theoretical framework devised by economists. Demonstrating that the network structure we observe actually has this or that economic implication is out of the scope of our work. It is the task of future research in economics to investigate to what extent the implications we assert can be verified. The existence of the core should be carefully analyzed and questioned in view of preserving global financial stability. In addition, is there any evidence that it ever acted as a bloc? Perhaps a very crucial outlook to be gained from our work is the question, to what extent policymakers and antitrust organizations should in future consider using international datasets and employ methods like ours to analyze the ensuing huge networks.

Moreover, the possible determinants of the observed structural organization should be discussed. To what extent can one find the micro-foundation which considers the interaction of economic agents not only amongst themselves but also with the market or the economy? Can the emergence of the economic "super-entity" be understood as a generalization of the rich club phenomenon (see Sect. 4.2.5) where now not the connectivity of the agents (i.e., the node degree) but their level of control plays an important role?

More specifically, our findings indicate novel questions relating to the study of firms. A systematic investigation could validate the conjecture that the relationship between the performance or profitability of a corporation and its position in the network has a functional relationship. Especially whether belonging to the "super-entity" provides a measurable competitive advantage. Another question concerns the notion of potential control of financial institutions. Further empirical work should be devoted to verifying to what extent they do actually exert control and why.

Moreover, ownership ties are often accompanied by other kinds of financial ties, such as credit or derivative contracts. As these are not disclosed for strategic reasons, it is an interesting perspective to see to what degree the ownership network can act as their proxy. Studies have suggested that financial ties are indeed correlated with ownership relations (Santos and Rumble 2006).

Regarding the empirical analysis of ownership networks, it would be highly illuminating to compare our study consisting of data from early 2007 with a current snapshot in order to observe and track the impact of the financial crisis. This generally opens the question of how ownership networks evolve. Next to the preliminary insights gained from the model of network formation, a data-driven, systematic analysis could uncover the real interplay between the network structure and its function in a dynamical setting. However, one would first need to address the issues of how to track and compare different ownership network snapshots in time.

The empirical observation of a bow-tie structure with a tiny but very influential core is novel in the study of complex networks, opening up the possibility that such a structure has been undetected in other real-world networks. We conjecture that it may be present in many different types of networks where rich-get-richer mechanisms are at work.

We are eager to know what insights the application of our 3-level methodology is able to reveal when applied to other weighted and directed complex networks with or without bow-tie topologies, where the nodes have an associated scalar value.

In general, the proposed Modeling Framework resulting in a network evolution yielding bow-tie topologies can easily be extended in many ways. Not only can the impact of different fitness values for the nodes be systematically investigated, in addition, the link-formation rules could be adapted to suit different contexts.

## 6.4 Contributions in a Nutshell

Our work aims at contributing to the understanding of economic systems from a real-world complex networks perspective. Focussing on the organization of national and global corporate control, for which economics does not offer any models or theories that predict its structure, our findings uncover previously unobserved properties and offer a theoretical model that explains the emergence of bow-tie structures in the presence of correlation. Our results are important for economics and the general study of complex networks, as summarized in the following.

### *6.4.1 Results for the Study of Complex Networks*

- The empirical studies of Chaps. 3 and 4 emphasize the importance of analyzing real-world networks using a 3-level approach,[10] reflected in the methodology given in Chap. 2. It is an interesting observation, that this line of approach is not yet well-established in the field. We believe that the future of network analysis lies in this direction and we hope our work contributes to the dissemination of this paradigm.
- Our methodology, although originating in economics, sheds new light on the notions of centrality and flow in networks, see Sects. 2.4 and 2.6. Notably, a novel centrality measure applicable for networks with bow-tie topologies is given in Eqs. (2.76) and (2.77).

[10] I.e., by considering the direction and weights of links next to assigning non-topological state variables to the nodes, see Sect. 1.1.1.

- The generic Modeling Framework given in Sect. 5.2.2 is able to produce arbitrary bow-tie components size in the network evolution.
- We give a general recipe for identifying and extracting the important nodes in complex networks: the backbone extraction algorithm of Sect. 3.4.
- The standard notions of degree are extended to encompass a 3-level network analysis, see Sect. 2.8.1.
- The observation that bow-tie topologies are not only present in technological networks but appear to be widespread in ownership networks. In particular, the first empirical evidence of a real-world complex network with a bow-tie topology where the SCC is tiny but comprised of the most important nodes.
- An overview of how the study of complex systems relates to the philosophy of science and can be understood in terms of formal models of reality, see Appendix A.

### *6.4.2 Results for Economics: Theoretical*

- A simple method to measure the concentration of a random variable is introduced in Sect. 2.10 .
- The existing limited models for estimating control from direct ownership relations are extended by a relative majority model, considering shareholder coalitions, see Sect. 2.8.
- The existing methodology for taking indirect ownership or control relations into account is refined by:
  - explicitly introducing the notions of integrated value (Sect. 2.3) and integrated control (Sect. 2.9), next to their corrections (Sect. 2.5);
  - uncovering the relationship between integrated value, network value and intrinsic value, and their control-related counterparts, as seen in Table 2.2;
  - implementing the corrections due to self-loops by identifying the correction operator, see Equation (2.50);
  - introducing the corrected network value and network control, see Sect. 2.5;
  - providing an algorithm to compute the control for large networks, see Sect. 2.5.2.
- The bow-tie model of ownership networks, described in Sect. 5.3, presents a first attempt at uncovering possible micro-interaction rules of economic agents yielding the observed macro patterns: the agents are maximizing their network control (i.e., their centrality).

### *6.4.3 Results for Economics: Empirical*

In a nutshell, the work presented in this thesis is the first massive empirical analysis of the global network of economic control. The results indicate that economic control is: (i) highly concentrated in the hands of few actors and much more concentrated than what was usually hypothesized by scholars and held in the public opinion; (ii) these powerful actors are not operating in isolation but are instead all interconnected in a tightly-knit group. Such a structure can align the interests of the most powerful actors and make them behave as a single economic "super-entity". These findings have never been previously reported, neither in the economic nor in the management literature.

In detail, the contributions are:

- Revealing that the US has the highest level of (national) concentration of market value, whereas CN and IT have the lowest, see Fig. 3.5.
- Uncovering that the more control is locally dispersed, the higher the global concentration of control, i.e., lying in the hands of very few important shareholders, and vice versa. This contrasts the idea of widely held firms in the US, see Chap. 3.
- Identifying the global key economic actors, as seen in Appendix E.
- Describing the topology of the global network of TNCs, especially the discovery of its bow-tie structure, see Sect. 4.2.
- Uncovering the surprisingly high concentration of corporate control world-wide, see Sect. 4.3.3.
- The identification of a tiny, highly interconnected, powerful economic "super-entity" comprised mainly of US corporations in the financial sector, see Sect. 4.3.5.
- Shedding a new light on the role of the financial sector, acting as a cohesive force by bridging communities and countries, see Sect. 4.2.7.

On a side note, the author's recent studies of foreign exchange markets resulted in the discovery of new empirical scaling laws (see Appendix C and Glattfelder et al. (2011) that substantially extend the catalogue of stylized facts. As these findings sharply constrain the space of possible theoretical explanations of the market mechanisms, this is an attempt at helping to understand the micro-workings of markets.

## References

W.A. Alley, Partial ownership arrangements and collusion in the automobile industry. J. Ind. Econ. **45**(2), 191–205 (1997)

H.V. Almeida, D. Wolfenzon, A theory of pyramidal ownership and family business groups. J. Finance **61**(6), 2637–2680 (2006)

H. Almeida, S.Y. Park, M. Subrahmanyam, D. Wolfenzon, Beyond cash flow and voting rights: valuation and performance of firms in complex ownership structures. Working paper, New York University Stern School of Business, 2007

E.S. Amundsen, L. Bergman, Will cross-ownership re-establish market power in the nordic power market? Energy J. **23**(2), 73–96 (2002)

D. Ariely, *Predictably Irrational: The Hidden Forces that Shape our Decisions* (Harper, New York, 2008)

S. Battiston, D. Delli Gatti, M. Gallegati, B. Greenwald, J.E. Stiglitz, credit chains and bankruptcy propagation in production networks. J. Econ. Dyn. Control **31**(6), 2061–2084 (2007)

S. Battiston, D.D. Gatti, M. Gallegati, B.C. Greenwald, J.E. Stiglitz, Liaisons dangereuses: increasing connectivity, risk sharing, and systemic risk. J. Econ. Dyn. Control **36**(8), 1121–1141 (2012a)

S. Battiston, M. Puliga, R. Kaushik, P. Tasca, G. Caldarelli. DebtRank: too central to fail? financial networks, the FED and systemic risk. Published online in Nature Scientific Reports on August 2, 2012b. http://www.nature.com/srep/2012/120802/srep00541/full/srep00541.html

J.P. Bouchaud, Economics needs a scientific revolution. Nature **455**(7217), 1181 (2008)

F. Brioschi, L. Buzzacchi, M.G. Colombo, Risk capital financing and the separation of ownership and control in business groups. J. Bank. Finance **13**(1), 747–772 (1989)

M. Buchanan, Waiting for the maths. Nat. Phys. **5**(11), 776 (2009)

A. Chapelle, Separation of ownership and control: where do we stand? Corp. Owners. Control **2**(2), 91–101 (2005)

A. Chapelle, A. Szafarz, Controlling firms through the majority voting rule. Physica A **355**, 509–529 (2005)

B. Chui, "Unified Theory" is Getting Closer, Hawking Predicts. San Jose Mercury News, 2000. http://www.comdig.com/stephen-hawking.php

S. Claessens, S. Djankov, The separation of ownership and control in east asian corporations. J. Finan. Econ. **58**(1–2), 81–112 (2000)

G.F. Davis, A new finance capitalism? mutual funds and ownership re-concentration in the united states. Eur. Manag. Rev. **5**(1), 11–21 (2008)

G.F. Davis, E.H. Kim, Business ties and proxy voting by mutual funds. J. Finan. Econ. **85**(2), 552–570 (2007)

E. Dietzenbacher, B. Smid, B. Volkerink, Horizontal integration in the dutch financial sector. Int. J. Ind. Organ. **18**(8), 1223–1242 (2000)

R. Dore, Stock market capitalism vs. Welfare Capitalism. New Polit. Econ. **7**(1), 115–127 (2002)

E.F. Fama, Efficient capital markets: a review of theory and empirical work. J. Finance **25**(2), 383–417 (1970)

E. Fama, Rational Irrationality. The New Yorker, 2010. http://www.newyorker.com/online/blogs/johncassidy/2010/01/interview-with-eugene-fama.html

J.D. Farmer, M. Gallegati, C. Hommes, A. Kirman, P. Ormerod, S. Cincotti, A. Sanchez, D. Helbing, A complex systems approach to constructing better models for managing financial markets and the economy, Submiss. Eur. Phys. J. (2012)

D. Gilo, Anticompetitive effect of passive investment. Michigan Law Rev. **99**(1) 1–47 (2000)

D. Gilo, Y. Moshe, Y. Spiegel, Partial cross ownership and tacit collusion. RAND J. Econ. **37**(1), 81–99 (2006)

J.B. Glattfelder, S. Battiston, Backbone of complex networks of corporations: the flow of control. Phys. Rev. E. **80**(3), 36104 (2009)

J.B. Glattfelder, A. Dupuis, R.B. Olsen, Patterns in high-frequency FX data: discovery of 12 empirical scaling laws. Quant. Finance **11**(4), 599–614 (2011)

A.G. Haldane, Rethinking the financial network. Speech at Financial Student Association, Amsterdam, 2009. http://www.bankofengland.co.uk/publications/speeches/2009/speech386.pdf

J.R. Iglesias, R.M.C de Almeida, Entropy and equilibrium state of free market models. Arxiv preprint arXiv:1108.5725, 2011. http://arxiv.org/abs/1108.5725

D. Kahneman, Maps of bounded rationality: psychology for behavioral economics. Am. Econ. Rev. **93**(5), 1449–1475 (2003)

P. Krugman, How Did Economists Get It So Wrong? The New York Times, 2009. http://www.nytimes.com/2009/09/06/magazine/06Economic-t.html

R. La Porta, F. Lopez de Silanes, A. Shleifer, R.W. Vishny, Law and finance. J. Polit. Econ. **106**(6), 1113–1155 (1998)

R. La Porta, F. Lopez de Silanes, A. Shleifer, Corporate ownership around the world. J. Finance **54**(2), 471–517 (1999)

J. Lorenz, S. Battiston, Systemic risk in a network fragility model analyzed with probability density evolution of persistent random walks. Netw. Heterog. Media **3**(2), 185 (2008)

T. Nenova, The value of corporate voting rights and control: a cross-country analysis. J. Finan. Econ. **68**(3), 325–351 (2003)
D.P. O'Brien, S.C. Salop, Competitive effects of partial ownership: financial interest and corporate control. Antitrust Law J. **67**, 559–614 (1999)
D.P. O'Brien, S.C. Salop, The competitive effects of passive minority equity interests: reply. Antitrust Law J. **69**, 611 (2001)
J.A.C. Santos, A.S. Rumble, The american keiretsu and universal banks: investing, voting and sitting on nonfinancials' corporate boards. J. Finan. Econ. **80**(2), 419–454 (2006)
J.E. Stiglitz, Risk and global economic architecture: Why Full Financial Integration May Be Undesirable. NBER Working Paper, 2010
J.E. Stiglitz, G. Akerlof, A new economics in an imperfect world. The Guardian, 2009. http://www.guardian.co.uk/commentisfree/2009/oct/28/economics-alternative-theories-stiglitz-regulators
S.H. Strogatz, Exploring complex networks. Nature **410**(6825), 268–276 (2001)
The deminor group. application of the one share—one vote principle in Europe. Technical report, A report prepared for Association of British Insurers by Deminor, 2005. http://www.abi.org.uk/Bookshop/
F. Trivieri, Does cross-ownership affect competition? evidence from the italian banking industry. J. Int. Finan. Markets Inst. Money **17**(1), 79–101 (2007)
S. Vitali, J.B. Glattfelder, S. Battiston, The network of global corporate control. PloS ONE **6**(10), e25995 (2011)
W. Wagner, Diversification at financial institutions and systemic crises. J. Finan. Intermed. **19**, 373–386 (2009)
M. Weber, *The Theory of Social and Economic Organization* (Free Press, New York, 1997)

# Appendix A
# Laws of Nature

*Why is there something rather than nothing?*

*For nothingness is simpler and easier than anything.*

(G.W. Leibniz in Leibniz 1954)

From all the conceivable questions regarding the foundation of reality most will probably remain forever in the domain of subjective speculation, metaphysical musings or theological postulations. One of the most basic questions to ask oneself is perhaps given in the quote above.
However, which questions have a scientific answer? A single question addressing many aspects of reality and science is:

**What are laws of nature?**

Or more specifically, what is science? In the history of science two schools of philosophy tried to tackle this question:

1. logical empiricism;
2. critical rationalism.

Naively one would expect science to adhere to the basic notions of common sense, logic, observations and experiments. Interestingly, these concepts turn out to be very problematic if applied to the question of what knowledge is and how it is acquired. After first looking in detail at these problems in the next section, the spectacular success of science and formal thought systems in describing fundamental and complex processes in nature is discussed in Sect. A.2.

## A.1 Ideas from the Philosophy of Science

The Greek philosopher Aristotle was one of the first thinkers to introduce logic as a means of reasoning. His empirical method was driven by gaining general insights from isolated observations. He had a huge influence on the thinking within the

J. B. Glattfelder, *Decoding Complexity*, Springer Theses,
DOI: 10.1007/978-3-642-33424-5, © Springer-Verlag Berlin Heidelberg 2013

Islamic and Jewish traditions next to shaping Western philosophy and inspiring the thinking in the physical sciences.

### A.1.1 Logical Empiricism

Nearly two thousand years later, not much had changed. Francis Bacon (the philosopher, not the painter) made modifications to Aristotle's ideas, introducing the so-called scientific method where inductive reasoning plays an important role. He paved the way for a modern understanding of scientific inquiry. Approximately at the same time, Robert Boyle was instrumental in establishing experiments as the cornerstone of physical sciences (around 1660).

By the early 20th century the notion that science is based on experience (empiricism) and logic, and where knowledge is intersubjectively testable, has had a long history. The philosophical school of *logical empiricism* (or logical positivism) tried to formalize these ideas. Notable proponents were Ernst Mach, Ludwig Wittgenstein, Bertrand Russell, Rudolf Carnap, Hans Reichenbach and Otto Neurath.

In this paradigm science is viewed as a building comprised of logical terms based on an empirical foundation. A theory is understood as having the following structure:

$$\text{observation} \rightarrow \text{empirical concepts} \rightarrow \text{formal notions} \rightarrow \text{abstract law}.$$

Basically a sequence of ever higher abstraction. This notion of unveiling laws of nature by starting with individual observations is called induction. The other way round, starting with abstract laws and ending with a tangible factual description is called deduction, see below.

What started off as a simple inquiry into the workings of nature soon faced serious difficulties. For instance:

- it turns out that it is not possible to construct pure formal concepts that solely reflect empirical facts without anticipating a theoretical framework;
- how does one link theoretical concepts (like electrons, utility functions in economics, inflational cosmology, Higgs bosons, etc.) to experiential notions?
- how to distinguish science from pseudo-science?

Now this may appear a little technical and not very interesting or fundamental to people outside the field of the philosophy of science, but the problems become more devastating: inductive reasoning is invalid from a formal logical point of view and causality defies standard logic (Brun and Kuenzle 2008).

### A.1.2 Critical Rationalism

The *critical rationalists* believed they could fix the problems the logical empiricists faced. A crucial influence came from René Descartes' and Gottfried Wilhelm Leibniz' rationalism: knowledge can have aspects that do not stem from experience, i.e., there is an immanent reality to the mind.

The term "critical" refers to the fact, that insights gained by pure thought cannot be strictly justified but only critically tested with experience. Ultimate justifications only lead to the so-called Münchhausen trilemma, i.e., one of the following:

1. an infinite regress of justifications;
2. circular reasoning;
3. dogmatic termination of reasoning.

The most influential proponent of critical rationalism was Karl Popper. His central claims were in essence to use deductive reasoning instead of induction and that theories can never be verified, only falsified.

Although there are similarities with logical empiricism (empirical basis, science as a set of theoretical constructs), the new idea is that theories are simply invented by the mind and are temporarily accepted until they can be falsified. The progression of science is hence seen as an evolutionary process rather than a linear accumulation of knowledge.

Unfortunately, also the school of critical rationalism faced unsurmountable challenges. In a nutshell:

- basic formal concepts cannot be derived from experience without induction and hence cannot be shown to be true;
- the notion deduction turns out to be just as tricky as that of induction;
- what parts of a theory need to be discarded once it is falsified?

To see where deduction breaks down, a nice story by Lewis Carroll (the mathematician who wrote the *Alice in Wonderland* stories): *What the Tortoise Said to Achilles.*[1]
If deduction goes down the drain as well, not much is left to ground science on notions of logic, rationality and objectivity. Which is rather unexpected of an enterprise that in itself works amazingly well employing just these concepts.
So what next? What are the consequences of these unexpected and spectacular failings of the most simplest premises one would wish science to be grounded on (logic, empiricism, causality, common sense, rationality, etc.)?

---

[1] See http://en.wikipedia.org/wiki/What_the_Tortoise_Said_to_Achilles.

### A.1.3 On the Horizon

In summary, some of the radical ideas scholars considered offering as new explanations.

**The Kuhnian View**

Thomas Kuhn's enormously influential work on the history of science is called the *Structure of Scientific Revolutions* (Kuhn 1962). He revised the idea that science is an incremental process accumulating more and more knowledge. Instead, he identified the following phases in the evolution of science:

- prehistory: many schools of thought coexist and controversies are abundant;
- history proper: one group of scientists establishes a new solution to an existing problem which opens the doors to further inquiry; a so called paradigm emerges;
- paradigm based science: unity in the scientific community on what the fundamental questions and central methods are; generally a problem solving process within the boundaries of unchallenged rules (analogy to solving a Sudoku);
- crisis: more and more anomalies and boundaries appear; questioning of established rules;
- revolution: a new theory and *weltbild* takes over solving the anomalies and a new paradigm is born.

Another central concept is *incommensurability*, meaning that proponents of different paradigms cannot understand the other's point of view because they have diverging ideas and views of the world. In other words, every rule is part of a paradigm and there exist no trans-paradigmatic rules.

This implies that such revolutions are not rational processes governed by insights and reason. In the words of Max Planck, found in his autobiography ((Planck 1950, pp. 33–34))

> *A new scientific truth does not triumph by convincing its opponents and making them see the light, but rather because its opponents eventually die, and a new generation grows up that is familiar with it.*

Kuhn gives additional blows to a commonsensical foundation of science with the help of Norwood Hanson and Willard Van Orman Quine:

- every human observation of reality contains an *a priori* theoretical framework;
- underdetermination of belief by evidence: any evidence collected for a specific claim is logically consistent with the falsity of the claim;
- every experiment is based on auxiliary hypotheses (initial conditions, proper functioning of apparatus, experimental setup, etc.).

People slowly started to realize that there are serious consequences in Kuhn's ideas and the problems faced by the logical empiricists and critical rationalists in establishing a sound logical and empirical foundation of science.

## Postmodernism

Modernism describes the development of the Western industrialized society since the beginning of the 19th century. Central ideas were that there exist objective true beliefs and that progression is always linear.

Postmodernism replaces these notions with the belief that many different opinions and forms can coexist and all find acceptance. Core ideas are diversity, differences and intermingling. In the 1970s it is seen to enter scientific and cultural thinking.

It is a notoriously hard concept to define, but the following quote by Richard Tarnas cuts to the heart of the matter, from Tarnas (1993):

> *There is an appreciation of the plasticity and constant change of reality and knowledge, a stress on the priority of concrete experience over fixed abstract principles, and a conviction that no single a priori thought system should govern belief or investigation. It is recognized that human knowledge is subjectively determined by a multitude of factors; that objective essences, or things-in-themselves, are neither accessible nor possible; and that the value of all truths and assumptions must be continually subjected to direct testing. The critical search for truth is constrained to be tolerant of ambiguity and pluralism, and its outcome will necessarily be knowledge that is relative and fallible rather than absolute or certain.*

Postmodernism has taken a bad rap from scientists after the so-called Sokal affair, where physicist Alan Sokal got a nonsensical paper published in the journal of postmodern cultural studies, by flattering the editor's ideology with nonsense that sounds good (Sokal 1996). Unfortunately, one could argue that similar strategies could generally work with scientific journals as well. Indeed, the Bogdanov Affair[2] is an academic dispute regarding the legitimacy of a series of theoretical physics papers written by French twins and TV personalities Igor and Grichka Bogdanov, that were published in reputable scientific journals. It was alleged that the contents was a meaningless combinations of buzzwords and the affair was covered in the mainstream media. The matter was also referred to as the "reverse Sokal" hoax. To this day the Bogdanov twins have insisted upon the validity of their work, however, the controversy has prompted reflections upon the peer-review system. On a more humorous note, the website http://www.snarXiv.org/ is aimed at spoofing the theoretical, high-energy physics section of the popular scientific archive for electronic preprints http://www.arXiv.org. The author even based an on-line game on this idea, where the player is offered two titles, an actual high-energy physics paper from the arXiv, and a completely fake title randomly generated by the snarXiv.

The notion of certainty can be seen as a central theme of postmodernism. The mathematician and philosopher Bertrand Russell is said to have lamented:

---

[2] See http://en.wikipedia.org/wiki/Bogdanov_Affair/.

*One of the painful things about our time is that those who feel certainty are stupid, and those with any imagination and understanding are filled with doubt and indecision.*

The origin of this dilemma can be traced back to the apparent feature of reality, that every explanation opens the door to new problems. As David Deutsch succinctly remarked in Deutsch (2011):

*The deeper an explanation is, the more new problems it creates.*

And Karl Popper's eloquent prose from Popper (1983):

*I think there is only one way to science—or to philosophy, for that matter: to meet a problem, to see its beauty and fall in love with it; to get married to it and to live with it happily, till death do ye part—unless you should meet another and even more fascinating problem or unless, indeed, you should obtain a solution. But even if you do obtain a solution you may then discover, to your delight, the existence of a whole family of enchanting, though perhaps difficult, problem children.*

In his widely acclaimed bestselling novel, which was rejected by over hundred publishers, Robert M. Pirsig addresses related philosophical issues. He argues that rational thought may find truth, but it may not be valid for all experiences. Moreover, the closer one examines a phenomenon, the more perplexing it becomes. Again, every explanation opens the door to countless new puzzles. See Pirsig (1974). Albert Einstein is attributed to having said, that there is no logical path to reach universal elementary laws of physics, only intuition can reach them. Finally, the behavioral economist Daniel Kahneman recently wrote in Kahneman (2011):

*Our comforting conviction that the world makes sense rests on a secure foundation: our near unlimited ability to ignore our ignorance.*

Postmodernisms has been associated with scepticism and solipsism, next to relativism and constructivism. Notable scientists identifiable as postmodernists are Thomas Kuhn, David Bohm and many figures in the 20th century philosophy of mathematics. As well as Paul Feyerabend, an influential philosopher of science.

## Constructivism

To quote the Nobel laureate Steven Weinberg on Kuhnian revolutions (Weinberg 1998):

*If the transition from one paradigm to another cannot be judged by any external standard, then perhaps it is culture rather than nature that dictates the content of scientific theories.*

*Constructivism* questions objectivism and rationality by postulating that beliefs are always subject to a person's cultural and theological embedding and inherent idiosyncrasies. In effect, constructed ideas. The notion also goes under the label of the *sociology of science*.

In its radical version, constructivism fully abandons objectivism and rationality:

> *Objectivity is the illusion that observations are made without an observer.* (Quote from the physicist Heinz von Foerster in Schülein and Reitze 2002, p. 174; translation mine.)

> *Modern physics has conquered domains that display an ontology that cannot be coherently captured or understood by human reasoning.* (Quote from the philosopher Ernst von Glasersfeld in Schülein and Reitze 2002, p. 175; translation mine.)

In addition, radical constructivism proposes that perception never yields an image of reality but is always a construction of sensory input and the memory capacity of an individual. An analogy would be the submarine captain who has to rely on instruments to indirectly gain knowledge from the outside world. Radical constructivists are motivated by modern insights gained by neuroscience, for instance (Eagleman 2011). Instead of reality being passively recorded by the brain, it is thought to be actively constructed by it, literally. Our brains sample just a small bit of the surrounding physical world and what we call normal perception does not really differ from hallucinations or dreams, except that the former is anchored by a little bit of external input. See also the ideas of the neurophysiologist Wolf Singer (Singer 2002), the cognitive scientist and philosopher Thomas Metzinger (Metzinger 2010) and the journalist and philosopher Richard Precht (Precht 2007).

Historically, Immanuel Kant can be understood as the founder of constructivism. On a side note, the bishop George Berkeley, an 18th century philosopher, went even as far as to deny the existence of an external material reality altogether. In summary, only ideas and thoughts are real. Knowledge about an object is what is perceived of it.

### Relativism

Another consequence of the foundations of science lacking commonsensical elements and the ideas of constructivism can be seen in the notion of relativism. If rationality is a function of our contingent and pragmatic reasons, then it can be rational for a group A to believe P, while at the same time it is rational for group B to believe in negation of P.

Although, as a philosophical idea, relativism goes back to the Greek Protagoras, its implications are unsettling for the Western mind: "anything goes" (as Paul Feyerabend characterizes his idea of scientific anarchy (Feyerabend 1975)). If there is no objective truth, no absolute values, nothing universal, then a great many of humanity's century old concepts and beliefs are in danger.

It should however also be mentioned, that relativism is prevalent in Eastern thought systems, and as an example found in many Indian religions. In a similar vein, pantheism and holism are notions which are much more compatible with Eastern thought systems than Western ones.

## In Summary

Consider the following issues.

- Epistemological:

  - problems with perception: synaesthesia, altered states of consciousness (spontaneous, mystical experiences and drug induced);
  - the field of psychopathology describes a frightening amount of defects in the perception of reality and ones self;
  - as an example, people suffering from psychosis or schizophrenia can experience a radically different reality;
  - the notion that consciousness and self-awareness is more akin to a virtual-reality rendering process (Singer 2002; Metzinger 2010);
  - free will appears problematic in neuroscience (Singer 2002; Metzinger 2010);
  - irrationalities uncovered in the studies of behavioral economists (Ariely 2008; Kahneman 2011);
  - the McGurk effect (http://bit.ly/McGurkEffectYT), or chronostasis;
  - the classic "rubber hand experiment", where an artificial hand is incorporated into one's body perception, see Botvinick et al. (1998), and its extension to the perceptual illusion of body swapping (Petkova and Ehrsson 2008);
  - beliefs, moral, ethics, trust and social behavior can all depend on what molecules are present in the brain at a certain time (Kosfeld et al. 2005; Crockett et al. 2008);
  - the idea of "synthetic happiness", the ability of the mind to construct experiences of happiness, by the psychologist Dan Gilbert (Gilbert 2006);
  - experience of pain and pleasure can depend on the context, for instance, tasting the same wine three times, but experiencing more pleasure if you thought it was expensive (Plassmann et al. 2008);
  - the Dunning-Kruger effect shows that unskilled individuals suffer from illusory superiority, mistakenly rating their ability much higher than average;
  - in general, the multitude of known cognitive biases.

- Ontological:

  - non-local foundation of quantum reality: entanglement;
  - illogical foundation of reality: wave-particle duality, superpositions, uncertainty, intrinsic probabilistic nature, observer/measurement problem in quantum theory;
  - discreteness of reality: quanta of energy and matter, constant speed of light;
  - nature of time: not present in fundamental theories of quantum gravity, symmetrical in all physical theories, depends on velocity and gravity (special and general relativity);
  - arrow of time: why was the initial state of the universe very low in entropy?
  - emergence, self-organization and structure-formation;
  - our universe appears fine-tuned, slightly changing the values of fundamental constants prohibits the formation of complex structures;
  - approximately 95 % of the contents of the universe is made up of some unknown substance (dark matter and dark energy);

- the apparent breakdown of causality, when quantum entanglement is paired with special relativity (Stefanov et al. 2003), or in delayed choice experiments (Ma et al. 2012).

In essence, perception doesn't necessarily say much about the world around us. Consciousness can fabricate reality. This makes it hard to be rational. Moreover, reality is quite a bizarre place. Only our macroscopic, classical level of reality appears to be well-behaved and so normal although it is based on intrinsic quantum weirdness.

And what about the human mind? Can we at least there find a paradox free realm? Unfortunately not. Even what appears as a consistent and logical formal thought system, i.e., mathematics, can be plagued by fundamental problems. Kurt Gödel proved that in every consistent non-contradictory system of mathematical axioms (leading to elementary arithmetics of whole numbers), there exist statements which cannot be proven or disproved in the system. So logical axiomatic systems are incomplete (Gödel 1931).

As an example Bertrand Russell encountered the following paradox: let $R$ be the set of all sets that do not contain themselves as members. Is $R$ an element of itself or not?

By only focussing on these foundational questions and problems one is driven to abandon any attempt in understanding reality. However, what is just as amazing as the discussed issues is the fantastic success of science and its unstoppable progression. This brings us back to the question posed at the beginning of this appendix: what are laws of nature?

## A.2 Laws of Nature: A Case Study

Given the issues discussed in the last section, it is an amazing feat that, firstly, there is any structure to reality and, secondly, that the mind can devise formal thought systems describing these regularities. Or, in other words, why has science been so fantastically successful at describing reality? And why is science producing the most amazing technology at breakneck speed?

At a recent workshop, from May 20–22, 2010, at the Perimeter Institute (for Theoretical Physics) in Waterloo, Ontario, Canada[3] these question were, once more, addressed: Why should nature be governed by laws? Why should those laws be expressible in terms of mathematics? The outcome was, unsurprisingly, still very ambiguous and intangible.[4]

[3] More details on their web-page http://www.perimeterinstitute.ca/Events/Laws_of_Nature/Laws_of_Nature:_Their_Nature_and_Knowability/.

[4] See also the Stanford encyclopedia of philosophy's article on what laws of nature are http://plato.stanford.edu/entries/laws-of-nature/.

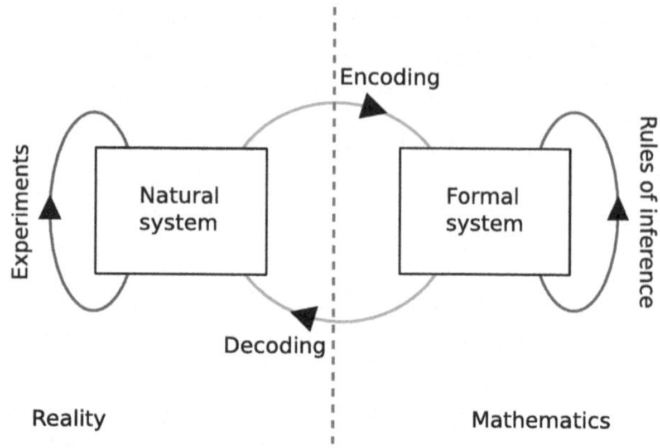


**Fig. A.1** Schematic illustration of the interplay between the outer reality and the mind's reality harboring formal thought systems

Without wanting to grapple any further with these notions and in being pragmatic, there are two domains of reality that have been described and understood with spectacular success by science: fundamental and complex processes, discussed in Sects. A.2.2 and A.2.3.
This is similar to the dilemma faced in quantum theory. On the one hand, all attempts to fully understand its notions and implications, hence giving an interpretation of the theory, have failed so far. But on the other hand, by simply using its mathematical framework as a tool, and by sweeping the conceptual problems under the rug, allows for the most fascinating technological advancements.

## *A.2.1 The Setting in a Nutshell*

Science is the quest to capture the processes of nature in formal mathematical representations. In other words, "mathematics is the blueprint of reality" in the sense that formal systems are the foundation of science. See Fig. A.1 for an overview of this notion, following (Casti 1989). Natural systems are a subset of reality, i.e., the observable universe. Guided by thought, observation and measurement natural systems are "encoded" into formal systems. Using logic (rules of inference) in the formal system, predictions about the natural system can be made (decoding). Checking the predictions with the experimental outcome gives the validity of the formal system as a model for the natural system.

Laws of nature can thus be understood as regularities and structures in a highly complex universe. They dependent on only a small set of conditions (i.e., are independent of very many conditions which could possibly have an effect).

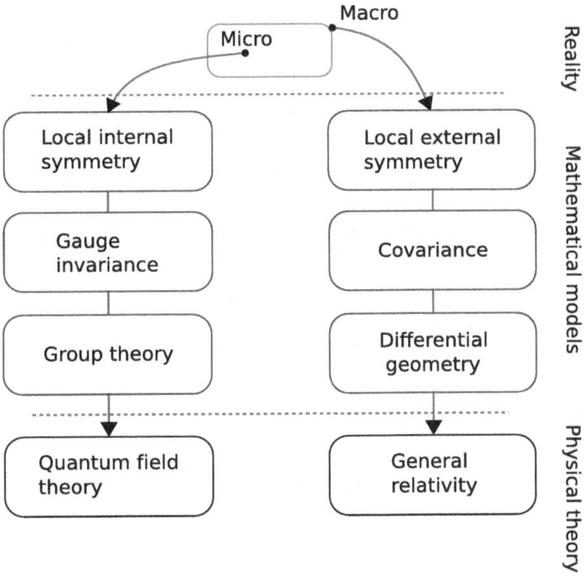


**Fig. A.2** A big chunk of reality described by the symmetry of mathematical models

## A.2.2 *The Paradigm of Fundamental Processes*

The paradigm in the study of fundamental processes in nature is:

**Mathematical models of reality are independent of their formal representation.**
This leads to the notions of symmetry and invariance. Basically, this requirement gives rise to nearly all of physics. See Fig. A.2 illustrating how a large part of physics can simply be understood from symmetry principles.

### Classical Mechanics

Symmetry, understood as the invariance of the equations under temporal and spacial transformations, gives rise to the conservation laws of energy, momentum and angular momentum.

In layman terms this means that the outcome of an experiment is unchanged by the time and location of the experiment and the motion of the experimental apparatus. Just a commonsense notion.

The intuitive notion of symmetry has been rigorously defined in the mathematical terms of group theory.

## Physics of Non-Gravitational Forces

The three non-gravitational forces are described in terms of quantum field theories. These in turn can be expressed as gauge theories, where the parameters of the gauge transformations are local, i.e., differ from point to point in space-time.

The Standard Model of elementary particle physics unites the quantum field theories describing the fundamental interactions of particles in terms of their (gauge) symmetries. See also the left-hand side of Fig. A.2.

## Physics of Gravity

Gravity is the only force that can't be expressed as a quantum field theory.
Its symmetry principle is called covariance, meaning that in the geometric language of the theory describing gravity (general relativity) the physical content of the equations is unchanged by the choice of the coordinate system used to represent the geometrical entities.

To illustrate, imagine an arrow located in space. It has a length and an orientation. In geometric terms this is a vector, lets call it $a$. If I want to compute the length of this arrow, I need to choose a coordinate system, which gives me the $x$-, $y$- and $z$-axes components of the vector, e.g., $a = (3, 5, 1)$. So starting from the origin of my coordinate system $(0, 0, 0)$, if I move 3 units in the x direction (left-right), 5 units in the y-direction (forwards-backwards) and 1 unit in the z direction (up-down), I reach the end of my arrow. The problem is now, that depending on the choice of coordinate system—meaning the orientation and the size of the units—the same arrow can look very different: $a = (3, 5, 1) = (0, 23.34, -17)$. However, every time I compute the length of the arrow in meters, I get the same number independent of the chosen representation.

In general relativity the vectors are somewhat like multidimensional equivalents called tensors and the commonsense requirement, that the calculations involving tensor do not depend on how they represent the tensors in space-time, is covariance. Again, see Fig. A.2.

It is quite amazing, but there is only one more ingredient needed in order to construct one of the most aesthetic and accurate theories in physics. It is called the equivalence principle and states that the gravitational force is equivalent to the forces experienced during acceleration. This may sound trivial, has however very deep implications.

## Physics of Condensed Matter

This branch of physics, also called solid-state physics, deals with the macroscopic physical properties of matter. It is one of physics first ventures into many-body problems in quantum theory. Although the employed notions of symmetry do not act at such a fundamental level as in the above mentioned theories, they are a cornerstone of the theory. Namely the complexity of the problems can be reduced using symmetry in order for analytical solutions to be found. Technically, the symmetry groups are boundary conditions of the Schrödinger equation. This leads to the theoretical framework describing, for example, semiconductors and so-called quasi-crystals (having fractal properties). In the super-conducting phase, the wave function becomes symmetric.

#### The Success

It is somewhat of a miracle that the formal systems the human brain discovers/devises find their match in the workings of nature. In fact, there is no reason for this to be the case, other than that it is the way things are.

E. Wigner captures this salient fact in his essay "The Unreasonable Effectiveness of Mathematics in the Natural Sciences" (Wigner 1960):

> *[...] the enormous usefulness of mathematics in the natural sciences is something bordering on the mysterious and [...] there is no rational explanation for it.*
>
> *[...] it is not at all natural that 'laws of nature' exist, much less that man is able to discover them.*
>
> *[...] the two miracles of the existence of laws of nature and of the human mind's capacity to divine them.*
>
> *[...] fundamentally, we do not know why our theories work so well.*

The following two examples should underline the power of this fact, where new features of reality where discovered solely on the requirements of the mathematical model. Firstly, in order to unify electromagnetism with the weak force (two of the three non-gravitational forces), the theory postulated two new elementary particles: the W and Z bosons. Needless to say, these particles where hitherto unknown and it took 10 years for technology to advance sufficiently in order to allow their discovery. Secondly, the fusion of quantum mechanics and special relativity lead to the Dirac equation which demands the existence of an, up to then, unknown flavor of matter: antimatter. Four years after the formulation of the theory, antimatter was experimentally discovered.

### *A.2.3 The Paradigm of Complex Systems*

While physics has had an amazing success in describing most of the observable universe in the last 300 years, the formalism appears to be restricted to the fundamental workings of nature. Only solid-state physics attempts to deal with collective systems. And only thanks to the magic of symmetry one is able to deduce fundamental analytical solutions.

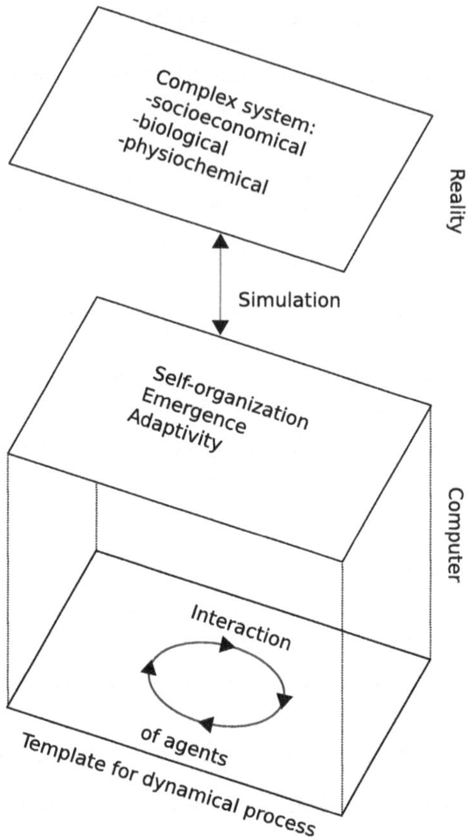


**Fig. A.3** Simulating a real-world complex system with an agent-based model

In order to approach real-life complex phenomena, one needs to adopt a more systems oriented focus. This also means that the interactions of entities becomes an integral part of the formalism.

Some ideas should illustrate the change in perspective:

- most calculations in physics are idealizations and neglect dissipative effects like friction;
- most calculations in physics deal with linear effect, as non-linearity is hard to tackle and is associated with chaos; however, most physical systems in nature are inherently non-linear (Strogatz 1994);
- the analytical solution of three gravitating bodies in classical mechanics, given their initial positions, masses, and velocities, cannot be found; it turns out to be a chaotic system which can only be simulated in a computer; however, there are an estimated hundred billion galaxies in the universe.

Figure A.5 at the end of this appendix shows a history of the various disciplines focussing on complex systems.
The study of complex systems appears hopelessly complicated, as it moves away from the reductionistic approach of established science. A quote from Anderson (1972) illustrates this fact:

> *At each stage [of complexity] entirely new laws, concepts, and generalizations are necessary [...]. Psychology is not applied biology, nor is biology applied chemistry.*

However, the paradigms of the study of complex systems are surprising:

1. **Every complex system is reduced to a set of objects and a set of functions between the objects.**
2. **Macroscopic complexity is the result of simple rules of interaction at the micro level.**

Paradigm 1 is reminiscent of the natural problem solving philosophy of object-oriented programming, where the objects are implementations of classes (collections of properties and functions) interacting via functions (public methods). A programming problem is analyzed in terms of objects and the nature of communication between them. When a program is executed, objects interact with each other by sending messages. The whole system obeys certain rules (encapsulation, inheritance, polymorphism, etc.).

Indeed, in mathematics the field of category theory defines a category as the most basic structure. It is as a set of objects and a set of morphisms (maps between the sets) (Hillman 2001). This resulted in a "unification of mathematics" in the 1940s.

A natural incarnation of a category is given by a complex network where the nodes represent the objects and the links describe their relationship or interaction. Now the structure of the network (i.e., the topology) determines the function of the network. See also Sect. 1.1.

Paradigm 2 is perhaps as puzzling as the "unreasonable effectiveness of mathematics in the natural sciences". To quote Stephen Wolfram's reaction to the realization that simplicity encodes complexity from (Wolfram (2002), p. 9):

> *And I realized, that I had seen a sign of a quite remarkable and unexpected phenomenon: that even from very simple programs behavior of great complexity could emerge.*
>
> *Indeed, even some of the very simplest programs that I looked at had behavior that was as complex as anything I had ever seen.*
> *It took me more than a decade to come to terms with this result, and to realize just how fundamental and far-reaching its consequences are.*

This also highlights the paradigm shift from mathematical (analytical) models to algorithmic models (computations and simulations performed in computers). In other words, the analytical description of complex systems is abandoned in favor of algorithms describing the interaction of the objects, called agents, in a system according to simple rules. This has given rise to the large field of *agent-based*

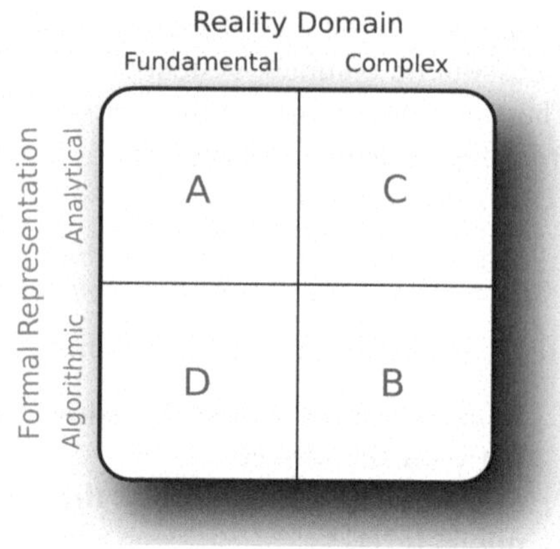


**Fig. A.4** Overview of the understanding of the laws of nature: domain of reality versus possible formal representations

*modeling*. Remarkably, this results in complex behavior: emergence, adaptivity, structure-formation and self-organization. In essence, complexity does not stem from the number of agents but the number of interactions. For instance there are roughly 25000 genes in a human versus about 50000 genes in a grain of rice.

Figure A.3 shows an illustration if this paradigm: in a computer simulation agents are interacting according to simple rules and give rise to patterns and behavior seen in real-world complex systems.

To be precise, there is still some mathematical formalism used in the study of complex systems. For instance, at the macro level, the so-called Fokker-Planck differential equation gives the collective time evolution of the probability density function of a system of agents as a function of time. While at the micro level, a single agent's behavior can be described by the so-called Langevin differential equation. The two formalism can be mapped into each other. However, as an example, 10000 agents following Langevin equations in a computer simulation approximate the macro dynamics of the system more efficiently than an ana-lytical investigation attempting to solve the equivalent Fokker-Planck differential equation.

### Success and Challenges

The modeling of animal swarming behavior, ant foraging, biological (temporal-spatial) pattern formation, population dynamics, pedestrian/traffic dynamics, market dynamics etc., which where hitherto impossible to tackle with a top-down approach, are well understood by the bottom-up approach of complex systems.

But:

- What are the right theoretical tools (methodology) to address complexity and are there unifying concepts?

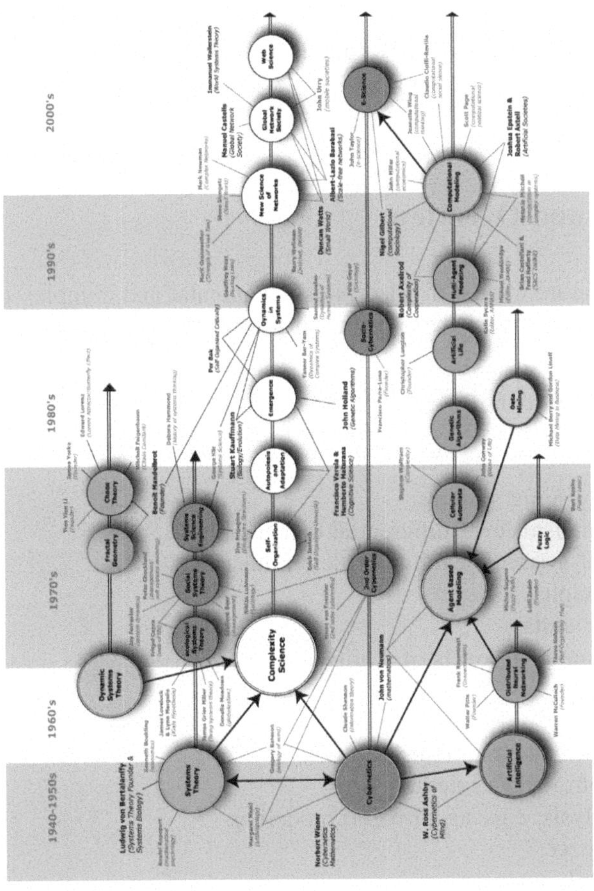


**Fig. A.5** A schematic illustration of the various fields dealing with the study of complex systems with a historical time-line; source Wikipedia (complexity map overview)

- How do the macroscopic system's properties depend on the microscopic interactions of the agents?
- How does one connect models to reality? How to make quantifiable, falsifiable predictions? How to empirically validation agent-based models?
- What are artifacts of the model? Is the model easier to understand than the physical system it is describing? Problems of parameter dependence and fine-tuning.

And finally, as the author's in Buchanan and Caldarelli (2010) observe:

> *One thing that is still missing is a complete theory of why nature is so fond of networks.*

**Are There Fundamental Laws of Nature in Complex Systems?**

Yes: scaling laws. See Appendix C.

## A.3 Summary and Overview

The workings of nature can be divided into fundamental and complex processes. The formal models describing these realms successfully have been analytical and algorithmic, respectively.

From Fig. A.4 we can identify:

- **A** (Sect. A.2.2)
  - Conservative dynamic systems.
  - Analytical formalism (e.g., differential equations).
  - Predictability of system's evolution (deterministic).
  - Symmetry principles.
  - Few interacting units.
  - Describes fundamental levels of reality.
- **B** (Sect. A.2.3)
  - Dissipative (real-world) systems.
  - Complexity from simple interactions.
  - Simulation analysis, e.g., agent-based modeling.
  - Algorithmic aspect.
  - Complex network analysis.
  - Unpredictability, stochasticity.
  - Systems can only be understood by letting them evolve.
  - Non-linear dynamics.
  - Many interacting units.
  - Describes socio-economic, biological and physio-chemical systems.
- **C** (some authors have recently argued that complex systems can and should be tackled with mathematical analysis (Sornette 2008))

- Right level of magnification of systems reveals order and organization.
- Partial predictability and pockets of predictability (meteorology, climate science, finance).

- **D** (mostly uncharted, but some tentative efforts are listed)

- Describing space-time as a network in some fundamental theories of quantum gravity (e.g., spin networks in loop quantum gravity).
- Deriving fundamental laws from cellular automaton networks (Wolfram 2002).
- Emergent complexity from fundamental quantum field theories (Täuber 2008).

## A.4 Outlook: The Computational Aspects of Reality

Information theory, especially quantum information, appears to be a very fundamental aspect of reality. Relativity is essentially about the transmission of information. Quantum theory is at heart about extracting information from reality: the information you receive depends on how you gather that information.

On the horizon, some scholars are slowly concluding that the most fundamental aspect of reality is in fact its computational capacities. To quote Seth Lloyd, a "quantum engineer" at the Massachusetts Institute of Technology:

> *The computational capability of the universe explains one of the great mysteries of nature: how complex systems such as living creatures can arise from fundamentally simple physical laws.*
> (S. Lloyd in (Lloyd 2006, p. 3))

Indeed, some notable scientists have alleged that reality *per se* is a computational process, i.e., quanta are in reality just information-processing bits (Zeilinger 2003) and the universe is similar to a computer (Lloyd 2006). For instance also John Archibald Wheeler in Wheele (1990)

> *'It from bit.' Otherwise put, every 'it'—every particle, every field of force, even the space-time continuum itself—derives its function, its meaning, its very existence entirely—even if in some contexts indirectly—from the apparatus-elicited answers to yes-or-no questions, binary choices, bits. 'It from bit' symbolizes the idea that every item of the physical world has at bottom—a very deep bottom, in most instances—an immaterial source and explanation; that which we call reality arises in the last analysis from the posing of yes–no questions and the registering of equipment-evoked responses; in short, that all things physical are information-theoretic in origin and that this is a participatory universe.*

The idea of a participatory epistemology, where meaning is derived at the focal point of interaction of the human mind with the world, goes back to Johann Wolfgang von Goethe and has been discussed extensively by the cultural historian

---

[5] Events run by local volunteers who are connected through the global Ignite network: http://igniteshow.com.

Richard Tarnas (Tarnas 1993):

> *Nature's reality is not merely phenomenal, nor is it independent and objective; rather, it is something that comes into being through the very act of human cognition. Nature becomes intelligible to itself through the human mind.*

In closing, and on a personal note, the contents of this chapter have been condensed to a 5-min talk I gave at an *Ignite event*[5] in Zurich: http://bit.ly/FabricatingReality.

# Appendix B
# Elements of Complex Network Theory

## B.1 Introduction

There are many excellent introductory texts, surveys, overviews and books covering the many topics related to complex networks: (Strogatz 2001; Albert and Barabási 2002; Dorogovtsev and Mendes 2002, 2003; Newman 2003; Newman et al. 2006; Caldarelli 2007; Costa et al. 2007.)

König and Battiston (2009) gives a graph-theoretic tutorial pertaining to economic networks.

In the following, some of the general concepts relating to this thesis are briefly introduced.

## B.2 Basic Notions

A network is a set of nodes connected by links. In graph theory,[6] networks are defined as graphs, links are referred to as edges and the nodes as vertices. Formally, a *graph* $G$ is a pair $G = (V, E)$, consisting of a set of vertices $V$ and a set of edges $E$, being a set of unordered pairs of elements $e_{ij}$ of $V$. The edge set of a loop-less graph constitutes an adjacency relation on $V$. Formally, an adjacency relation is any relation which is anti-reflexive and symmetric. In other words, if two vertices $i$ and $j$ are adjacent, then $e_{ij} \in E$.

The *adjacency matrix* $A = A(G)$ of a graph is a matrix with rows and columns labeled by graph vertices, where $A_{ij} = 1$ if the edge $e_{ij}$ exists, otherwise $A_{ij} = 0$. For an undirected graph $A$ is symmetric. If the graph is weighted, the entry $A_{ij} = W_{ij}$ represents the edge from $i$ to $j$ with the weight $W_{ij}$.

---

[6] See for instance West (2001).

J. B. Glattfelder, *Decoding Complexity*, Springer Theses,
DOI: 10.1007/978-3-642-33424-5, © Springer-Verlag Berlin Heidelberg 2013

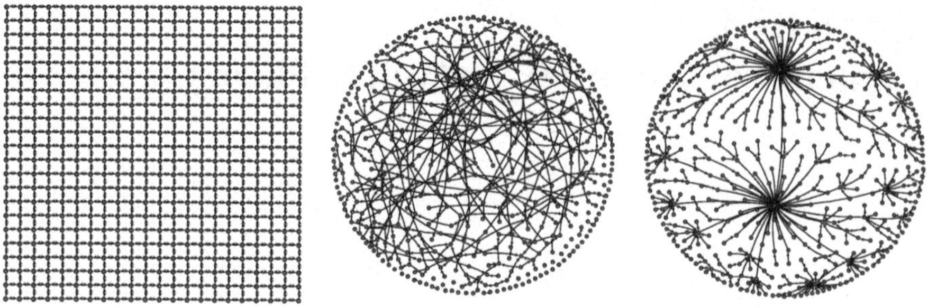

**Fig. B.1** Examples of common network topologies: (*left*) a regular two-dimensional lattice, (*middle*) a random network with an average degree of one, (*right*) a scale-free network with an average degree of one showing two hubs; reproduced with kind permission from Geipel (2010)

The number $k_i$ of edges per vertex $i$ is called the *degree*. If the edges are oriented, one has to distinguish between the in-degree and out-degree, $k^{in}$ and $k^{out}$, respectively. When the edges $ij$ are weighted with the number $W_{ij}$, the corresponding quantity is called *strength* (Barrat et al. 2004):

$$k_i^w := \sum_j W_{ij}. \tag{B.1}$$

Note that for weighted and oriented networks, one has to distinguish between the in- and out-strengths, $k^{in-w}$ and $k^{out-w}$, respectively.

## B.3 Network Topologies

The network topology specifies the physical (i.e., real) or logical (i.e., virtual) arrangement of the elements of a network. The topology can be considered as the shape or structure of a network.

Random network topologies were first studied in Rényi and Erdős (1960). They are discussed in Sect. B.6 and represent the opposite of regular networks, see Fig. B.1.

Central to this thesis are so-called bow-tie topologies, described in Sect. 1.1.1.

The discovery of two very special network topologies, present in many real-world complex networks, sparked the initial interest in the study of complex networks, detailed in the following. See also Buchanan and Caldarelli (2010) for a discussion of the impact and popularity of such networks in science.

### B.3.1 Scale-Free Networks

The degree distribution of most complex networks follows a scaling-law probability distribution $\mathcal{P}(k) \propto x^{-\alpha}$, see also Barabási and Albert (1999); Albert and Barabási (2002); Caldarelli (2007). Scaling laws are explained in Appendix C. Scale-free networks are characterized by high robustness against random failure of nodes, but susceptible to coordinated attacks on the hubs. Theoretically, they are thought to arise from a dynamical growth process, called preferential attachment, in which new nodes favor linking to existing nodes with high degree (Barabási and Albert 1999). Although alternative mechanisms have been proposed (Caldarelli et al. 2002).

In detail, the degree sequence of $k$, $k^{in}$ or $k^{out}$ follows a scaling-law probability distribution with exponent $\alpha$ if

$$\mathcal{P}(k) = \frac{k^{-\alpha} e^{-k/\kappa}}{Li_{\alpha}(e^{-1/\kappa})}. \tag{B.2}$$

The exponential term in the numerator, governed by the parameter $\kappa$, results in an exponential cutoff, the term in the denominator ensures the proper normalization, and $Li_n(x)$ is the $n$th polylogarithm of $x$ (Newman et al. 2001; Albert and Barabási 2002). Note that for the limit $\kappa \to \infty$

$$\mathcal{P}(k) = \frac{k^{-\alpha}}{\zeta(\alpha)}, \tag{B.3}$$

where the Riemann $\zeta$-function now functions as the normalization constant.

### B.3.2 Small-World Networks

Watts and Strogatz (1998) introduced the notion of "small-world" networks. In these networks, although most nodes are not neighbors of one another, all nodes can be reached from every other by a surprisingly small number of hops or steps. The psychologist Milgram popularized this notion in 1967 with an experiment (Milgram 1967). The phrase "six degrees of separation" is associated with the experiments' outcome, implying the idea that everyone is at most six steps away from any other person on Earth.

As a modern example, Twitter,[7] a microblogging service less than a few years old, commanded more than 41 million users as of July 2009 and is growing fast, with over 140 million users early in 2012. The average path length on Twitter was found to be 4.12 (Kwak et al. 2010).

---

[7] URL http://www.twitter.com.

Formally, small-world networks are characterized by a high clustering coefficient[8] and a low average path length. They can be understood to lie between regular lattices and random networks. The presence of hubs is responsible for the high connectivity of the network, see Fig. B.1. Small-world networks are constructed using link rewiring or link addition (Watts and Strogatz 1998; Newman and Watts 1999).

Observe that although scale-free networks are also small-world networks, the opposite is not always true. However, many real-world complex networks show both scale-free and small-world characteristics.

## B.4 Perron-Frobenius Theorem

The *eigenvalues* of the adjacency matrix $A$ are the numbers $\lambda$ such that $Ax = \lambda x$ has a nonzero solution vector, which is an *eigenvector* associated with $\lambda$.

The term $\lambda_{\text{PF}}$ denotes the largest real eigenvalue of $A$, the *Perron-Frobenius eigenvalue*. For all other eigenvalues $\lambda$ of $A$ one finds $|\lambda| \leq \lambda_{\text{PF}}$ and there exists an associated nonnegative eigenvector $v \geq 0$ such that $Av = \lambda_{\text{PF}} v$.

For a connected graph $G$ the adjacency matrix $A(G)$ has a unique largest real eigenvalue $\lambda_{\text{PF}}$ and a positive associated eigenvector $v > 0$. See also Horn and Johnson (1990); Seneta (2006).

## B.5 Community Analysis

A community (also called cohesive group) is a subset of nodes with dense interconnections among its members, and sparse relations with nodes located in other communities. Researchers have proposed many different methods for community detection (Clauset et al. 2004; Donetti and Munoz 2004; Capocci et al. 2005; Newman 2006; Reichardt and Bornholdt 2006; Alves 2007). However, many of them have some limitations. In fact, only a few algorithms are able: (i) to process big networks in a reasonable time and (ii) to return good partitions without the need to specify in advance the number of the final communities or their size. See Newman (2004) for a review. All these algorithms can be categorized as divisive, agglomerative and optimization methods.

The method we apply in Sect. 4.2.7 belongs to the latter class (Blondel et al. 2008). Briefly, the quality of the partitions is commonly measured by the so-called "modularity" (i.e., a function that measures the goodness of a particular partition

[8] A measure of the degree to which nodes in a graph tend to cluster together, derived from the number of triangles present in the network.

of the network), which is also used as objective function to be optimized. Such methods have the advantage of finding high modularity partitions in large networks, in short time, by unfolding a complete hierarchical community structure and without arbitrary *ex ante* assumptions.

## B.6 Random Networks

### *B.6.1 Undirected Random Networks*

A random undirected graph consists of $n$ nodes, $l$ links and link probability $p$, with binomial degree distribution

$$\mathcal{P}(k_i = k) = \binom{n}{k} p^k (1-p)^{n-k}, \tag{B.4}$$

where $k_i$ is the degree of node $i$ (Rényi and Erdős 1960; Albert and Barabási 2002; Newman 2003; Park and Newman 2004). The first terms gives the number of equivalent choices of such a network. The remaining terms describe the probability of a graph with $k$ links and $n$ nodes existing.

Note that a network of $n$ nodes has maximally the following number of links

$$l_{max} = \frac{1}{2} n(n-1), \tag{B.5}$$

and it holds that

$$z := \langle k \rangle = \frac{l}{n} = p(n-1). \tag{B.6}$$

In the limit of large $n$ the following approximations become exact

$$\mathcal{P}(k) \approx \frac{z^k e^{-z}}{k!}, \tag{B.7}$$

$$z \approx pn. \tag{B.8}$$

This can be seen by noting that

$$e^{-z} = \lim_{n \to \infty} \left( 1 + \frac{-z}{n} \right), \tag{B.9}$$

$$1 = \lim_{n \to \infty} \left( \frac{n!}{n^k (n-k)!} \right). \tag{B.10}$$

Observe that Eq. (B.7) describes a Poisson distribution.

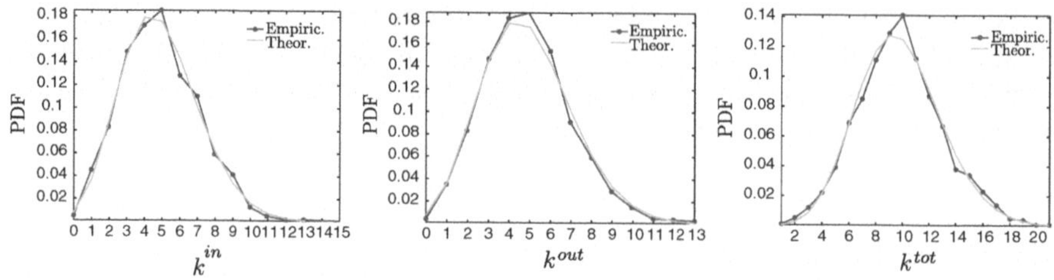


**Fig. B.2** Theoretical degree distributions, defined in Eq. (B.12), of a directed random network with $p_{tot} = 0.01$ compared to an empirical realization ($n = 1000$, $l = 4900$): (*left*) $k_{in}$ with $\langle z_{in} \rangle = 4.9$, (*middle*) $k_{out}$ with $\langle z_{out} \rangle = 4.9$, (*right*) $k_{tot} = k_{in} + k_{out}$ with $\langle z_{tot} \rangle = 9.8$

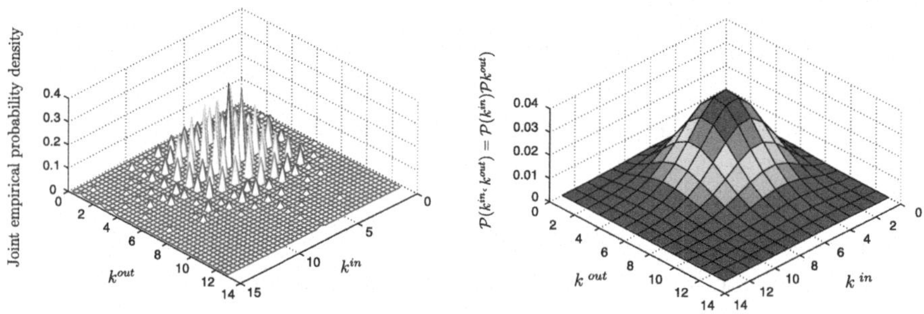


**Fig. B.3** Joint degree distributions of the same network analyzed in Fig. B.2: (*left*) empirical realization, (*right*) theoretical values given by Eq. (B.12)

## B.6.2 Directed Random Networks

Directed networks are described by a joint probability density $\mathcal{P}(k_{in}, k_{out})$. Note, however, that generally this function is not factorizable, i.e., $\mathcal{P}(k_{in}, k_{out}) \neq \mathcal{P}(k_{in})\mathcal{P}(k_{out})$.

The realization of the random directed network is based on the undirected case, where the links get assigned a direction with equal probability. This choice splits the number of links seen by each node to $k_{tot} := k_{in} + k_{out}$ and $k_{in} = k_{out}$. One finds in this case that

$$\mathcal{P}(k_{in}, k_{out}) = \mathcal{P}(k_{in})\mathcal{P}(k_{out}), \tag{B.11}$$

$$\mathcal{P}(k_*) = \frac{{z_*}^{k_*} e^{-z_*}}{k_*!}, \quad * = in, out, tot, \tag{B.12}$$

$$2z_{in} = 2z_{out} = z_{tot} = p_{tot} n = (p_{in} + p_{out})n, \tag{B.13}$$

where $p_{tot}$ corresponds to $p$ of the undirected case, $p_{in} = p_{out} = 0.5 p_{tot}$ and $z = z_{tot}$. Figures. B.2 and B.3 compare these theoretical values to realizations of random networks.

The relationship between the degree distributions of the directed and undirected case are given by Dorogovtsev et al. (2001)

$$\mathcal{P}(k) = \sum_{k_{in}} \mathcal{P}(k_{in}, k - k_{in}). \tag{B.14}$$

In the case of a directed random graph, Eqs. (B.11) and (B.12), one finds

$$\sum_{k_{in}} \mathcal{P}(k_{in}, k - k_{in}) = \sum_{k_{in}} \frac{{z_{in}}^{k_{in}} e^{-z_{in}}}{k_{in}!} \frac{\langle k - k_{in}\rangle^{k-k_{in}} e^{\langle k-k_{in}\rangle}}{(k - k_{in})!} \tag{B.15}$$

$$\overset{\langle k-k_{in}\rangle = z_{in}}{=} z_{in}^{k} e^{-z_{tot}} \sum_{k_{in}} \frac{1}{k_{in}!(k - k_{in})!} \tag{B.16}$$

$$= \frac{{z_{tot}}^{k} e^{-z_{tot}}}{k!} = \frac{z^k e^{-z}}{k!} = \mathcal{P}(k), \tag{B.17}$$

by noting that

$$2^k = \sum_{i} \frac{k!}{i!(k - i)!}. \tag{B.18}$$

## B.7 Pagerank

Pagerank is a link analysis algorithm used by the Google Internet search engine. It assigns a numerical weighting to each element of a hyperlinked set of documents, such as the World-Wide Web (WWW), with the purpose of measuring its relative importance within the set (Brin and Page 1998). It can be understood as a centrality measure, where a node's importance is given by the importance of the nodes pointing to it.

Pagerank is formally defined by an iterative equation

$$pr_i(t+1) = \alpha \sum_{j \in \Gamma(i)} \frac{pr_j(t)}{k_j^{out}} + \frac{1-\alpha}{N}, \tag{B.19}$$

where $\Gamma(i)$ is the set of labels of the neighboring nodes of $i$. Or in matrix notation

$$pr(t+1) = \alpha \mathcal{M} pr(t) + \frac{1-\alpha}{N}\mathbf{1}, \tag{B.20}$$

where $\mathbf{1}$ is the unit column-vector and the matrix $\mathcal{M}$ is

$$\mathcal{M}_{ij} = \begin{cases} 1/k_j^{out}, & \text{if } j \text{ links to } i; \\ 0, & \text{otherwise.} \end{cases} \tag{B.21}$$

Alternatively, $\mathcal{M} := (K^{-1}A)^t$, if $K$ is the diagonal matrix with the out-degrees in the diagonal and $A$ is the adjacency matrix of the network.

The solution is given, in the steady state, by

$$pr = (\mathbf{I} - \alpha\mathcal{M})^{-1}\frac{\mathbf{1} - \alpha}{\mathbf{N}}\mathbf{1}, \tag{B.22}$$

with the identity matrix **I**. It exists and is unique for $0 < \alpha < 1$, where $\alpha$ is called the dampening factor and is usually set to 0.85.

Conceptually, the formula reflects a model of a random surfer in the WWW who gets bored after several clicks and switches to a random page. The Pagerank value of a page measures the chance that the random surfer will land on that page by clicking on a link. If a page has no links to other pages, it becomes a sink and therefore terminates the random surfing process, unless $\alpha < 1$. In this case, the random surfer arriving at a sink page, jumps to a random web page chosen uniformly at random. Hence $(1 - \alpha)/N$ in Eqs. (B.19) and (B.20) is interpreted as a teleportation term.

# Appendix C
# Scaling Laws

Scaling-law relations characterize an immense number of natural processes, prominently in the form of

1. scaling-law distributions;
2. scale-free networks, see Sect. B.3.1;
3. cumulative relations of stochastic processes.

A scaling law, or power law, is a simple polynomial functional relationship

$$f(x) \quad \propto \quad x^{-\alpha}. \tag{C.1}$$

Two properties of such laws can easily be shown:

- a logarithmic mapping yields a linear relationship;
- scaling the function's argument $x$ preserves the shape of the function $f(x)$, called scale invariance.

See for instance Newman (2005), Sornette (2000).

## C.1 Scaling-Law Distributions

Scaling-law distributions have been observed in an extraordinary wide range of natural phenomena: from physics, biology, earth and planetary sciences, economics and finance, computer science and demography to the social sciences (West et al. 1997; Amaral et al. 1998; Albert et al. 1999; Sornette 2000; Pastor-Satorras et al. 2001; Bouchaud 2001; Newman et al. 2002; Caldarelli et al. 2002; Garlaschelli et al. 2003; Gabaix et al. 2003; Newman 2005; Lux 2008; Di Matteo 2007).

It is truly amazing, that such diverse topics as

- the size of earthquakes, moon craters, solar flares, computer files, sand particle, wars and price moves in financial markets;

J. B. Glattfelder, *Decoding Complexity*, Springer Theses,
DOI: 10.1007/978-3-642-33424-5, © Springer-Verlag Berlin Heidelberg 2013

- the number of scientific papers written, citations received by publications, hits on web-pages and species in biological taxa;
- the sales of music, books and other commodities;
- the population of cities;
- the income of people;
- the frequency of words used in human languages and of occurrences of personal names;
- the areas burnt in forest fires;

are all described by scaling-law distributions. First used to describe the observed income distribution of households by the economist Pareto in 1897 (Pareto 1897), the recent advancements in the study of complex systems have helped uncover some of the possible mechanisms behind this universal law. However, there is still no conclusive understanding of the origins of scaling laws. Some insights have been gained from the study of critical phenomena and phase transitions, stochastic processes, rich-get-richer mechanisms and so-called self-organized criticality (Bouchaud 2001; Barndorff-Nielsen and Prause 2001; Farmer and Lillo 2004; Newman 2005).

Processes following normal distributions have a characteristic scale given by the mean of the distribution. In contrast, scaling-law distributions lack such a preferred scale. Measurements of scaling-law processes yield values distributed across an enormous dynamic range (sometimes many orders of magnitude), and for any section one looks at, the proportion of small to large events is the same. Historically, the observation of scale-free or self-similar behavior in the changes of cotton prices was the starting point for Mandelbrot's research leading to the discovery of fractal geometry (Mandelbrot 1963).

It should be noted, that although scaling laws imply that small occurrences are extremely common, whereas large instances are quite rare, these large events occur nevertheless much more frequently compared to a normal (or Gaussian) probability distribution. For such distributions, events that deviate from the mean by, e.g., 10 standard deviations (called "10-sigma events") are practically impossible to observe. For scaling law distributions, extreme events have a small but very real probability of occurring. This fact is summed up by saying that the distribution has a "fat tail" (in the terminology of probability theory and statistics, distributions with fat tails are said to be leptokurtic or to display positive kurtosis) which greatly impacts the risk assessment. So although most earthquakes, price moves in financial markets, intensities of solar flares, ...will be very small, the possibility that a catastrophic event will happen cannot be neglected.

## C.2 Cumulative Scaling-Law Relations

Next to distributions of random variables, scaling laws also appear in collections of random variables, called stochastic processes. Prominent empirical examples are financial time-series, where one finds empirical scaling laws governing the relationship between various observed quantities. See Müller et al. (1990), Mantegna and Stanley (1995), Guillaume et al. (1997), Galluccio et al. (1997), Dacorogna and Gencay (2001), Glattfelder et al. (2011).

# Appendix D
# Some Technical Aspects

## D.1 Proving That the Algorithmic Methodology Corrects for Cycles

Here we now show that the breadth-first-search (BFS) algorithm presented in Sect. 2.5.2 yields an equivalent computation proposed in the literature to address the problems of the presence of cycles leading to exaggerated network value.

In Brioschi et al. (1989) the notion of network value was introduced based on ownership

$$v^{net} = \widetilde{W}v + v, \tag{D.1}$$

which in Baldone et al. (1998) was identified as being problematic. The authors hence introduced a new model which overcomes this problem of exaggerated indirect value in presence of cycles. This is given by

$$\hat{v}^{net} = \widehat{W}v + \mathcal{D}v, \tag{D.2}$$

where the matrix $W$ is corrected by removing the links which from $i$ indirectly coming back to $i$:

$$\widehat{W}_{ij} = W_{ij} + \sum_{k \neq i} \widehat{W}_{ik} W_{kj} \tag{D.3}$$

and

$$\mathcal{D} := \text{diag}\big((I - W)^{-1}\big)^{-1} = I - \text{diag}(\widehat{W}). \tag{D.4}$$

Our proposed algorithmic solution also corrects for cycles in an equivalent way. This can be seen as follows. By applying the BFS algorithm to node $i$, we extract the adjacency matrix $B(i)$ of the subnetwork of nodes downstream. From Eq. (D.3) it holds by construction that

J. B. Glattfelder, *Decoding Complexity*, Springer Theses,
DOI: 10.1007/978-3-642-33424-5, © Springer-Verlag Berlin Heidelberg 2013

$$\tilde{B}(i)_{ij} = \widehat{W}_{ij} - \widehat{W}_{ii}, \tag{D.5}$$

where $\tilde{B}(i)$ is defined equivalently to Eq. (2.15). In a more compact notation

$$\tilde{B}(i)_{i*} = \widehat{W}_{i*} - [\text{diag}(\widehat{W})]_{i*}. \tag{D.6}$$

Employing Eq. (D.4) we find that $\tilde{B}(i)_{i*} + I_{i*} = \widehat{W}_{i*} + \mathcal{D}_{i*}$, or equivalently

$$\widehat{W}_{i*}v + \mathcal{D}_{i*}v = \mathcal{D}_{i*}(\widetilde{W}_{i*}v + v_i) = \mathcal{D}_{i*}v^{net} = \hat{v}_i^{net} \tag{D.7}$$

$$= \tilde{B}_{i*}(i)v + v_i = \tilde{v}^{int}(i) + v_i = v^{net}(i). \tag{D.8}$$

This concludes that our algorithmic method and the results in Baldone et al. (1998) are identical: $\hat{v}_i^{net} = v^{net}(i)$, recalling Eq. (2.59).

In summary, this means that for any node $i$ in a strongly connected component, the algorithmic BFS method computes the loop-corrected network value. As mentioned in Sects. 2.3.5 and 2.4.1, there is a second problem related to root nodes accumulating too much value in comparison to other nodes. This problem is fixed in Step 4 of the algorithm detailed on p. 45, which deals with each component of the bow-tie topology specifically.

For the integrated value in this context, see end of Sect. 2.5.3.

## D.2 The Relationship Between the Degree and the Fraction of Control

In Sect. 2.8 the relative model of control $C^{RM}$ was defined in Eq. (2.125), respectively Eq. (2.123), employing the fraction of control $H_{ij}$. It was then claimed in Sect. 2.8.2 that this measure is similar to the so-called degree of control $\alpha$. This quantity was introduced as a probabilistic voting model measuring the degree of control of a block of large shareholdings as the probability of it attracting majority support in a voting game.

In Leech (1987) the degree of control of the largest shareholder is approximated as

$$\alpha_1 \approx \Phi_{0,1}\left(\frac{w_1}{\sqrt{\mathcal{H} - w_1^2}}\right), \tag{D.9}$$

where $\Phi$ is the cumulative normal distribution function, $\mathcal{H}$ the Herfindahl index, defined in Eq. (2.121), and

$$\Phi_{\mu,\sigma^2}(x) := \frac{1}{\sigma\sqrt{2\pi}} \int_{-\infty}^{x} e^{-\frac{(u-\mu)^2}{2\sigma^2}} du = \frac{1}{2}\left[1 + \text{erf}\left(\frac{x-\mu}{\sigma\sqrt{2}}\right)\right]. \tag{D.10}$$

Considering the stock labelled with 1 having $n$ shareholders and the largest one also being called 1, i.e., $W_{11}$ is the biggest shareholding. Hence Eq. (2.123) yields

$$H_{i1} = \frac{W_{i1}^2}{\sum_{l=1}^{k_1^{in}} W_{l1}^2} = \frac{W_{i1}^2}{\mathcal{H}}. \tag{D.11}$$

Thus

$$\alpha_1 \approx \Phi_{0,1}\left(\frac{W_{11}}{\sqrt{\mathcal{H} - W_{11}^2}}\right) = \Phi_{0,\sigma^2}(W_{11}), \tag{D.12}$$

where $\sigma^2 = \mathcal{H} - W_{11}^2$. This yields

$$\alpha_1 \approx \frac{1}{\sqrt{2\pi}} \int_{-\infty}^{W_{11}/\sqrt{\mathcal{H}-W_{11}^2}} e^{-\frac{u^2}{2}} du. \tag{D.13}$$

A weak conjecture is that there is a linear relation between the two quantities, i.e.,

$$\alpha_1 \approx a + b \cdot H_{11}. \tag{D.14}$$

A stronger conjecture is that the two functions

$$\alpha_1 \approx \Phi_{0,\sigma^2}(W_{11}), \tag{D.15}$$

and

$$H_{11} = \frac{W_{11}^2}{\sigma^2 + W_{11}^2}, \tag{D.16}$$

behave in a similar way. To approach these questions, in the following, numerical examples are given.

## *Empirical Examples*

In Fig. D.1 the $\alpha_1$ and $H_{11}$ coordinates are drawn for the following distributions of percentages of shareholdings:

1. 0.8, 0.01, 0.011, 0.08, 0.05, 0.012;
2. 0.25, 0.25, 0.25, 0.25;
3. 0.49, 0.3, 0.1, 0.1;
4. 0.49, 0.02, 0.019, 0.015, 0.015, 0.013, 0.012, 0.01;
5. 0.3, 0.3, 0.1;
6. 0.2, 0.2, 0.2, 0.05, 0.02, 0.014;
7. 0.1, 0.1, 0.1, 0.015, 0.015, 0.013, 0.012, 0.01;
8. 0.012, 0.011, 0.014, 0.015, 0.015, 0.013, 0.012, 0.01;

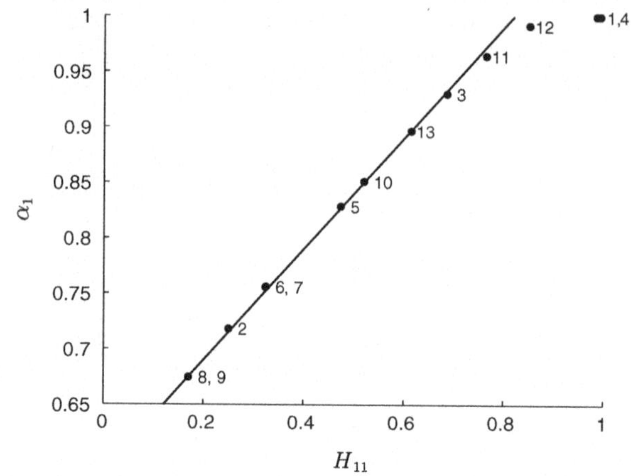


**Fig. D.1** Correlation between $\alpha_1$ and $H_{11}$ for 13 shareholding configurations mentioned in the text; the straight line is given by $y = 0.59 + 0.5x$

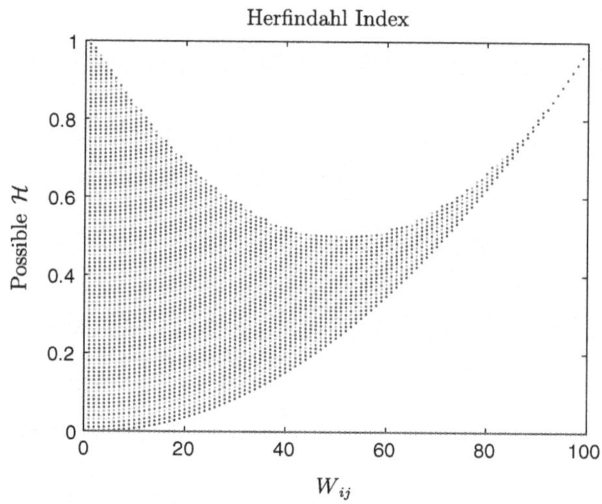


**Fig. D.2** The Herfindahl index $\mathcal{H}$ as a function of $W_{ij}$

9. 0.12, 0.11, 0.14, 0.15, 0.15, 0.13, 0.12, 0.1;
10. 0.51, 0.49;
11. 0.51, 0.20, 0.20;
12. 0.51, 0.1, 0.11, 0.08, 0.05, 0.12;
13. 0.3, 0.15, 0.14, 0.12.

There is a prominent linear relationship seen for most data points in Fig. D.1, with the exceptions of points (1), (4) and (12) which all characterize situations in which there is one very dominant shareholder and very highly dispersed ownership for the remaining shareholders. This means that $W_{11} \gg W_{ij}$, $(i,j) \neq (1,1)$ and $\mathcal{H} \approx 1$. Observe that the Herfindahl index $\mathcal{H}$ is a function of the ownership weights $W_{ij}$. Namely, for every $W_{ij}$ the possible range of $\mathcal{H}$ is given by $[W_{ij}^2, W_{ij}^2 + (1 - W_{ij})^2]$. This is plotted in Fig. D.2.

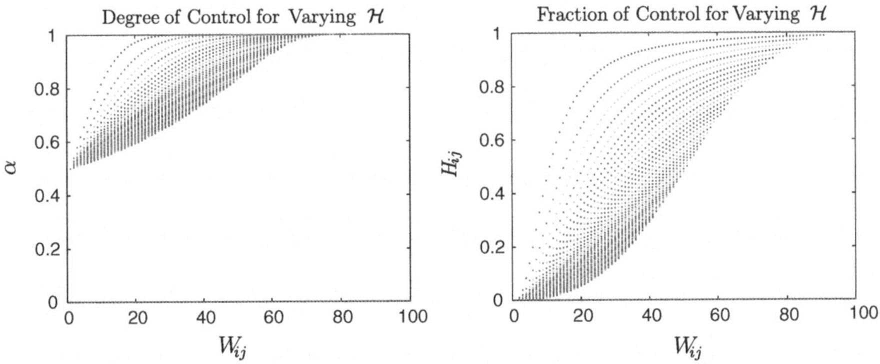


**Fig. D.3** $\alpha$ (*left*) and $H_{ij}$ (right) as functions of $W_{ij}$ for different configurations of $\mathcal{H}$

In Fig. D.3, the values of $\alpha$ and $H_{ij}$ are shown for the varying values of $\mathcal{H}$ defined by $W_{ij}$. There is a striking agreement in the qualitative behavior of the two measures. Indeed, $H_{ij}$ emulates the behavior of $\alpha$ on the full $y$-axis range, from 0 to 100 %.

The discrepancy mentioned above can be understood by looking at the behavior of the top lines, representing large $\mathcal{H}$ values. For these curves $\alpha$ approaches the value of one faster than $H_{ij}$. This concludes that there is a striking qualitative agreement for the two measures of control. In other words, the integral expression of Eq. (D.13) can, after proper rescaling of the co-domain, be represented as the simple fraction given in Eq. (D.16).

# Appendix E
# Who are the Global Key Economic Actors?

In Chap. 1 two reoccurring questions were:

> Who are the key economic actors holding the largest fraction of control?
>
> To what degree are the top economic actors interconnected with each other?

By performing a level-3 network analysis of the global ownership network of TNCs (see Chap. 4; a list of acronyms can be found in front matter of this book.), it has become, for the first time, possible to answer these questions. Only the bird's-eye-view given by a network perspective can uncover the true organization of key economic actors.

In Table E.1 the top 50 corporate power-holders are listed. The importance of the economic actors is related to their level of integrated control, as seen in Sects. 2.5.2 and 2.5.3. The threshold model for estimating direct control, introduced in Eq. (2.118), is chosen. The intrinsic value of the TNCs is given by their operating revenue. The full list can be found here: http://ethz.focproject.net/viewer/tnc.

Not only is the cumulative increase in the percentage of integrated control shown, tracking the incremental decrease in importance of the top power-holders, crucially their position in the network is also shown. From this it is apparent that the top economic actors are highly interconnected and organize as an economic "super-entity" in the SCC of the global network of corporations, see also Sect. 4.3.5. Another interesting observation is that most of the top power-holders are financial intermediaries.

See Sect. 6.1 for a discussion of the relevance of these findings.

J. B. Glattfelder, *Decoding Complexity*, Springer Theses,
DOI: 10.1007/978-3-642-33424-5, © Springer-Verlag Berlin Heidelberg 2013

**Table E.1** List of the first 50 corporate power-holders with country, industrial sector (NACE) and network position information; the list is ordered by integrated control $\mathring{\varsigma}^{\text{int}}$ (TM)

| Rank | Economic actor name | Country | NACE code | Network position | Cumulative integrated control (TM, %) |
|---|---|---|---|---|---|
| 1 | BARCLAYS PLC | GB | 6512 | SCC | 4.05 |
| 2 | CAPITAL GROUP COMPANIES INC, THE | US | 6713 | IN | 6.66 |
| 3 | FMR CORP | US | 6713 | IN | 8.94 |
| 4 | AXA | FR | 6712 | SCC | 11.21 |
| 5 | STATE STREET CORPORATION | US | 6713 | SCC | 13.02 |
| 6 | JPMORGAN CHASE & CO. | US | 6512 | SCC | 14.55 |
| 7 | LEGAL & GENERAL GROUP PLC | GB | 6603 | SCC | 16.02 |
| 8 | VANGUARD GROUP, INC., THE | US | 7415 | IN | 17.25 |
| 9 | UBS AG | CH | 6512 | SCC | 18.46 |
| 10 | MERRILL LYNCH & CO., INC. | US | 6712 | SCC | 19.45 |
| 11 | WELLINGTON MANAGEMENT CO. L.L.P. | US | 6713 | IN | 20.33 |
| 12 | DEUTSCHE BANK AG | DE | 6512 | SCC | 21.17 |
| 13 | FRANKLIN RESOURCES, INC. | US | 6512 | SCC | 21.99 |
| 14 | CREDIT SUISSE GROUP | CH | 6512 | SCC | 22.81 |
| 15 | WALTON ENTERPRISES LLC | US | 2923 | T&T | 23.56 |
| 16 | BANK OF NEW YORK MELLON CORP. | US | 6512 | IN | 24.28 |
| 17 | NATIXIS | FR | 6512 | SCC | 24.98 |
| 18 | GOLDMAN SACHS GROUP, INC., THE | US | 6712 | SCC | 25.64 |
| 19 | T. ROWE PRICE GROUP, INC. | US | 6713 | SCC | 26.29 |
| 20 | LEGG MASON, INC. | US | 6712 | SCC | 26.92 |
| 21 | MORGAN STANLEY | US | 6712 | SCC | 27.56 |
| 22 | MITSUBISHI UFJ FINANCIAL GROUP, INC. | JP | 6512 | SCC | 28.16 |
| 23 | NORTHERN TRUST CORPORATION | US | 6512 | SCC | 28.72 |
| 24 | SOCIÉTÉ GÉNÉRALE | FR | 6512 | SCC | 29.26 |
| 25 | BANK OF AMERICA CORPORATION | US | 6512 | SCC | 29.79 |
| 26 | LLOYDS TSB GROUP PLC | GB | 6512 | SCC | 30.30 |
| 27 | INVESCO PLC | GB | 6523 | SCC | 30.82 |
| 28 | ALLIANZ SE | DE | 7415 | SCC | 31.32 |
| 29 | TIAA | US | 6601 | IN | 32.24 |
| 30 | OLD MUTUAL PUBLIC LIMITED COMPANY | GB | 6601 | SCC | 32.69 |
| 31 | AVIVA PLC | GB | 6601 | SCC | 33.14 |
| 32 | SCHRODERS PLC | GB | 6712 | SCC | 33.57 |
| 33 | DODGE & COX | US | 7415 | IN | 34.00 |
| 34 | LEHMAN BROTHERS HOLDINGS, INC. | US | 6712 | SCC | 34.43 |

(continued)

**Table E.1** (continued)

| Rank | Economic actor name | Country | NACE code | Network position | Cumulative integrated control (TM, %) |
|---|---|---|---|---|---|
| 35 | SUN LIFE FINANCIAL, INC. | CA | 6601 | SCC | 34.82 |
| 36 | STANDARD LIFE PLC | GB | 6601 | SCC | 35.2 |
| 37 | CNCE | FR | 6512 | SCC | 35.57 |
| 38 | NOMURA HOLDINGS, INC. | JP | 6512 | SCC | 35.92 |
| 39 | THE DEPOSITORY TRUST COMPANY | US | 6512 | IN | 36.28 |
| 40 | MASSACHUSETTS MUTUAL LIFE INSUR. | US | 6601 | IN | 36.63 |
| 41 | ING GROEP N.V. | NL | 6603 | SCC | 36.96 |
| 42 | BRANDES INVESTMENT PARTNERS, L.P. | US | 6713 | IN | 37.29 |
| 43 | UNICREDITO ITALIANO SPA | IT | 6512 | SCC | 37.61 |
| 44 | DEPOSIT INSURANCE CORPORATION OF JP | JP | 6511 | IN | 37.93 |
| 45 | VERENIGING AEGON | NL | 6512 | IN | 38.25 |
| 46 | BNP PARIBAS | FR | 6512 | SCC | 38.56 |
| 47 | AFFILIATED MANAGERS GROUP, INC. | US | 6713 | SCC | 38.88 |
| 48 | RESONA HOLDINGS, INC. | JP | 6512 | SCC | 39.18 |
| 49 | CAPITAL GROUP INTERNATIONAL, INC. | US | 7414 | IN | 39.48 |
| 50 | CHINA PETROCHEMICAL GROUP CO. | CN | 6511 | T&T | 39.78 |

# Appendix F
# Media Coverage

## F.1 In the Spotlight

### *F.1.1 The Backbone of Control*

When the study of Chap. 3 was first uploaded onto the arXiv platform,[9] it was covered in their blog in February 2009. The title of the post was "Econophysicists Identify World's Top 10 Most Powerful Companies".[10] This sparked a wave of interest in the blogosphere. Subsequent coverage included the blog of the Financial Times.[11]

Science News did an online article, titled "Networks Reveal Concentrated Ownership of Corporations" based on an interview with S. Battiston, see Sect. F.3.3. The study also got some international coverage by the national Austrian Broadcasting agency ORF.[12]

After the work was accepted for publication in August 2009,[13] LiveScience contacted us for an interview which lead to the online article given in F.3.4: "World's Stocks Controlled by Select Few". This triggered another wave of interest in the work. USA Today covered the story in their blog.[14]

---

9 An electronic archive of scientific preprints, see http://www.arxiv.org.

10 http://arxivblog.com/?p=1195.

11 "Who Controls the Stock Market, or, Physicists do it Differently", http://ftalphaville.ft.com/blog/2009/02/17/52539/who-controls-the-stock-market-or-physicists-do-it-differently/.

12 See http://sciencev1.orf.at/science/news/154497

13 In September the publication appeared, Chap. 3.

14 "Tiny Group of Stockholders Control Most Markets" http://content.usatoday.com/communities/sciencefair/post/2009/08/68497787/1.

J. B. Glattfelder, *Decoding Complexity*, Springer Theses,
DOI: 10.1007/978-3-642-33424-5, © Springer-Verlag Berlin Heidelberg 2013

### *F.1.2 Global Corporate Control*

In the fall of 2011, the study of Chap. 4 was prominently covered in the New Scientist article "Revealed: the capitalist network that runs the world", see Sect. F.3.1. The article happened to be published the week that the Occupy Wall Street movement became a global phenomenon. This coincidence of events resulted in the widespread international dissemination of our study: it went viral. For weeks, scientists, activists, journalists, radio hosts, even documentary filmers, from all over the world, tried contacting us. Our study was discussed intensely and controversially in the blogosphere. Unfortunately many people did not appreciate the fact, that we simply uncovered new findings in empirical data, without a political agenda or wanting to promote conspiracy theories. Some of the reaction were particularly vicious. For instance, Yves Smith from the blog "Naked Capitalism", who already disagreed with our 2009 study, wrote a scathing sensationalist critique and chose to ignore all subsequent clarifications and rebuttals from our side.[15]

In order to provide a larger context, offer general explanations and to stress our neutral position in the matter, I decided to write this blog post: http://j-node.blogspot.ch/2011/10/network-of-global-corporate-control.html. However, many people did agree about the relevance of our work, as a first glimpse into the organizing principles of a highly interconnected but uncharted economic system. Andy Haldane, executive director of financial stability at the Bank of England in London, welcomed our study, because in his eyes it represents a leap forward, see Sect. F.3.2.

For a discussion of some common misconceptions and the relevance of our findings, consult Sect. 6.1.

## F.2 The Other Side of the Coin...

The publicity of the 2009 study also caught the attention of a special-interest audience: people adhering to conspiracy theories. The empirical inequality in the distribution of control was seen as proof of the existence of an elite group controlling the world. Some of the titles circulating the Internet include "Illuminati Proven by Physics", "Physicists Shed Light on Illuminati" and "New World Order, Interlocking Directorships".

Indeed, the study appeared to spur the exuberant imagination of some people. One author saw a connection between our findings and solar cycles, purporting that the sun is the hidden influence behind global finance in a recently published book.[16]

---

[15] http://bit.ly/NakedCapitalismControlStudy.

[16] "The Hidden System: How the Sun Governs the Financial World and Influences Stock Prices" (translation mine): http://www.scorpio-verlag.de/default.asp?Menue=14&Buch=7.

## F.3 Archive

A summary of some notable media coverage is given in the following.

### F.3.1 New Scientist: Revealed—The Capitalist Network that Runs the World

*AS PROTESTS against financial power sweep the world this week, science may have confirmed the protesters' worst fears. An analysis of the relationships between 43,000 transnational corporations has identified a relatively small group of companies, mainly banks, with disproportionate power over the global economy.*

*The study's assumptions have attracted some criticism, but complex systems analysts contacted by New Scientist say it is a unique effort to untangle control in the global economy. Pushing the analysis further, they say, could help to identify ways of making global capitalism more stable.*

*The idea that a few bankers control a large chunk of the global economy might not seem like news to New York's Occupy Wall Street movement and protesters elsewhere (see photo). But the study, by a trio of complex systems theorists at the Swiss Federal Institute of Technology in Zurich, is the first to go beyond ideology to empirically identify such a network of power. It combines the mathematics long used to model natural systems with comprehensive corporate data to map ownership among the world's transnational corporations (TNCs).*

*"Reality is so complex, we must move away from dogma, whether it's conspiracy theories or free-market," says James Glattfelder. "Our analysis is reality-based." Previous studies have found that a few TNCs own large chunks of the world's economy, but they included only a limited number of companies and omitted indirect ownerships, so could not say how this affected the global economy—whether it made it more or less stable, for instance.*

*The Zurich team can. From Orbis 2007, a database listing 37 million companies and investors worldwide, they pulled out all 43,060 TNCs and the share ownerships linking them. Then they constructed a model of which companies controlled others through shareholding networks, coupled with each company's operating revenues, to map the structure of economic power.*

*The work, to be published in PLoS One, revealed a core of 1318 companies with interlocking ownerships (see image). Each of the 1318 had ties to two or more other companies, and on average they were connected to 20. What's more, although they represented 20 per cent of global operating revenues, the 1318 appeared to collectively own through their shares the majority of the world's large blue chip and manufacturing firms—the "real" economy—representing a further 60 per cent of global revenues.*

*When the team further untangled the web of ownership, it found much of it tracked back to a "super-entity" of 147 even more tightly knit companies—all of their ownership was held by other members of the super-entity—that controlled 40 per cent of the total wealth in the network. "In effect, less than 1 per cent of the companies were able to control 40 per cent of the entire network," says Glattfelder. Most were financial institutions. The top 20 included Barclays Bank, JPMorgan Chase & Co, and The Goldman Sachs Group.*

*John Driffill of the University of London, a macroeconomics expert, says the value of the analysis is not just to see if a small number of people controls the global economy, but rather its insights into economic stability.*

*Concentration of power is not good or bad in itself, says the Zurich team, but the core's tight interconnections could be. As the world learned in 2008, such networks are unstable. "If one [company] suffers distress," says Glattfelder, "this propagates."*

*"It's disconcerting to see how connected things really are," agrees George Sugihara of the Scripps Institution of Oceanography in La Jolla, California, a complex systems expert who has advised Deutsche Bank.*
*Yaneer Bar-Yam, head of the New England Complex Systems Institute (NECSI), warns that the analysis assumes ownership equates to control, which is not always true. Most company shares are held by fund managers who may or may not control what the companies they part-own actually do. The impact of this on the system's behaviour, he says, requires more analysis.*

*Crucially, by identifying the architecture of global economic power, the analysis could help make it more stable. By finding the vulnerable aspects of the system, economists can suggest measures to prevent future collapses spreading through the entire economy. Glattfelder says we may need global anti-trust rules, which now exist only at national level, to limit over-connection among TNCs. Sugihara says the analysis suggests one possible solution: firms should be taxed for excess interconnectivity to discourage this risk.*

*One thing won't chime with some of the protesters' claims: the super-entity is unlikely to be the intentional result of a conspiracy to rule the world. "Such structures are common in nature," says Sugihara.*
*Newcomers to any network connect preferentially to highly connected members. TNCs buy shares in each other for business reasons, not for world domination. If connectedness clusters, so does wealth, says Dan Braha of NECSI: in similar models, money flows towards the most highly connected members. The Zurich study, says Sugihara, "is strong evidence that simple rules governing TNCs give rise spontaneously to highly connected groups". Or as Braha puts it: "The Occupy Wall Street claim that 1 per cent of people have most of the wealth reflects a logical phase of the self-organising economy."*

*So, the super-entity may not result from conspiracy. The real question, says the Zurich team, is whether it can exert concerted political power. Driffill feels 147 is too many to sustain collusion. Braha suspects they will compete in the market but act together on common interests. Resisting changes to the network structure may be one such common interest.*

By Andy Coghlan, New Scientist issue 2835, 24 October 2011.[17]

## F.3.2 New Scientist: To Navigate Economic Storms We Need Better Forecasting

*When it comes to handling complexity, finance could learn a lot from meteorology. Time for action*

*IT IS said that economists engage in financial forecasting to make astrologers look good. So it is with some trepidation that I want to venture into a little financial futurology.*

*This act of bravery was prompted by a recent piece of research about the structure of the global corporate network that was reported in New Scientist (22 October, p 8). This gave us a tantalising glimpse of a brave new world for finance. I want to try and map that world. If we do so, I believe fragilities in the global financial system which have stalked the globe for four years (and counting) could be significantly reduced.*

*Finance is a classic complex, adaptive system, similar to an ecosystem. The growth in its scale, complexity and adaptation in the past generation alone would rival that of most other complex systems in the past century. The growth in certain financial markets and instruments has far outstripped Moore's Law of a doubling of computer power every 18 months. The stock of outstanding financial contracts is now around 14 times annual global GDP.*

*Yet this dense cat's cradle of finance has been woven largely out of sight. At best we are able to snatch passing glimpses of it, for data are incomplete, local and lagging. Making sense of the financial system is more an act of archaeology than futurology.*

*How so? Because, at least historically, finance has not thought of itself as a system. Instead, financial theory, regulation and data-collection has tended to focus on individual firms. Joining the dots was in no one's job description.*

*Theorists were partly to blame. Economics has always been desperate to burnish its scientific credentials and this meant grounding it in the decisions of individual people. By itself, that was not the mistake. The mistake came in thinking the behaviour of the system was just an aggregated version of the behaviour of the individual. Almost by definition, complex systems do not behave like this.*

---

[17] The article can be found online at: http://www.newscientist.com/article/mg21228354.500-revealed–the-capitalist-network-that-runs-the-world.html.

*Interactions between agents are what matters. And the key to that is to explore the underlying architecture of the network, not the behaviour of any one node. To make an analogy, you cannot understand the brain by focusing on a neuron—and then simply multiplying by 100 billion.*

*Regulators and statisticians were no less culpable. Pre-crisis, they sampled the financial neurons ever more intensively, but largely ignored the configuration of the financial cortex. When parts started to malfunction, this meant no one had much idea what critical faculties would be impaired.*

*That uncertainty, coupled with dense financial wiring, turned small failures into systemic collapse. In financial terms, Lehman Brothers was a small fish. But through network uncertainty and complex wiring, it polluted the whole financial pond.*

*Those experiences are now seared onto the conscience of regulators. Systemic risk has entered their lexicon, and to understand that risk, they readily acknowledge the need to join the dots across the network. So far, so good. Still lacking are the data and models necessary to turn this good intent into action.*

*There are grounds for optimism. Other disciplines have cut a dash in their complex network mapping over the past generation, assisted by increases in data-capture and modelling capability made possible by technology. One such is weather forecasting, which involves collecting over 100 million pieces of raw data daily from across the globe and using it to map evolving weather systems in near-real time and then simulating forward.*

*Success stories can also be told about utility grids and transport networks, the web, social networks, global supply chains and perhaps the most complex web of all, the brain.*

*Finance, by contrast, has been a late adopter. Analysis such as that on "the network that runs the world" is welcome because it represents a leap forward. A key ingredient for success elsewhere has been a common language and shared access to data. At present, finance has neither. But that, too, is set to change. Regulators are talking seriously about introducing common metrics for financial transactions. Alongside that, data warehouses are being constructed to store these raw materials.*

*There are even moves afoot to put these raw materials to work. The US aims to create an Office of Financial Research to collect data from firms and weave the information into a web suitable for mapping and simulating risk.*

*So to play clairvoyant, imagine the scene a generation hence. There is a single nerve centre for global finance. Inside, a map of financial flows is being drawn in real time. The world's regulatory forecasters sit monitoring the financial world, perhaps even broadcasting it to the world's media.*

*National regulators may only be interested in a quite narrow subset of the data for the institutions for which they have responsibility. These data could be part of, or distinct from, the global architecture.*

*Now imagine the light this financial map might shine. It would allow regulators to issue the equivalent of weather-warnings—storms brewing over Lehman Brothers, credit default swaps and Greece. It would enable advice to be issued—keep a safe distance from Bear Stearns, sub-prime mortgages and Icelandic banks. And it would enable "what-if?" simulations to be run—if UK bank Northern Rock is the first domino, what will be the next?*

*At present, global finance does not have the technology to ask these questions, much less answer them. The prize is a big, but attainable, one. It would not entirely lift the fog from finance, but it could provide us with a navigation system better able to avoid the next crash landing.*

By Andy Haldane, New Scientist issue 2842, 13 December 2011.[18]

## F.3.3 Science News: Networks Reveal Concentrated Ownership of Corporations

### Analysis of Stock Markets in 48 Countries Finds Backbones of Control

*Researchers have made the first maps of corporate stock ownership for the stock markets of a large number of countries, 48 in all. The new network analysis technique reveals "backbones" in these ownership networks: big players that together own a controlling stake in more than 80 percent of the companies in the markets.*

*In these network diagrams, nodes represent either a company with publicly owned stock or a shareholder. Links between the nodes show which shareholders hold stock in which companies. Because many publicly owned companies also hold shares in other companies, many nodes have both "owner" and "ownee" links. Plotting all these connections creates a map of the ownership structure of a stock market.*

*Unlike the approach used in the new study, simpler network analyses can't reveal these backbones of ownership because the market values of companies being traded aren't taken into account. The new study, published online February 5 at arXiv.org, adds these market values. It also includes a way to account for indirect ownership, such as when a company owns stock in a second company that, in turn, owns stock in a third company.*

[18] The article can be found online at:http://www.newscientist.com/article/mg21228425.400-to-navigate-economic-storms-we-need-better-forecasting.html.

*"If you do a network analysis, you can see things that you couldn't see otherwise," says Stefano Battiston, coauthor of the study and a physicist at the Swiss Federal Institute of Technology in Zurich who studies complex socioeconomic networks. "Although from an individual point of view corporations are widely held, from a global point of view ownership is more highly concentrated."*

*The resulting networks, which are based on a snapshot of market data from early 2007, show that concentration of ownership in these markets varies from country to country. The United States and United Kingdom had the highest concentration of ownership, while ownership was less concentrated in European and Asian countries. Some companies held so much stock at the time that they constituted part of the backbones of many countries. The top ten such companies were:*

1. *The Capital Group Companies (U.S.)*
2. *Fidelity Management & Research (U.S.)*
3. *Barclays PLC (U.K.)*
4. *Franklin Resources (U.S.)*
5. *AXA (France)*
6. *JPMorgan Chase & Co. (U.S.)*
7. *Dimensional Fund Advisors (U.S.)*
8. *Merrill Lynch & Co. (U.S.)*
9. *Wellington Management Company (U.S.)*
10. *UBS (Switzerland)*

*"The results nicely show how structure emerges from an otherwise weak signal, revealing the ownership backbone within and across countries," comments Bruce Kogut, an economist at the Columbia Business School in New York.*

*Most of these companies manage mutual funds, so they hold large portfolios of a wide variety of stocks on behalf of their clients. It's not surprising then that they would top the list, but the new study confirms this intuition with hard data.*

*"It's interesting that you can get these results that, if you asked an experienced economist they'd probably have a gut feeling about, but now you can show it in a quantitative way," comments Jörg Reichardt, a physicist at the University of Wuerzburg in Germany. "They've done a great job of making it mathematically rigorous."*

*The implications of these backbones of concentrated ownership for the current global economic crisis are unclear, Battiston says. He and his colleagues are now analyzing stock market data from after the economic downturn for comparison and working on theories for how the structure of ownership affects the overall stability of the market.*

*"In contrast to the mainstream economic view that a more interconnected market is always more stable, in many cases it can actually be more unstable because there are some mechanisms that have not been accounted for in the economic*

*theory so far," Battiston says. "If there are amplification systems and feedback, then a more connected world is more unstable."*

By Patrick Barry, Science News web edition, 13 February 2009.[19]

## *F.3.4 LiveScience: World's Stocks Controlled by Select Few*

*WASHINGTON—A recent analysis of the 2007 financial markets of 48 countries has revealed that the world's finances are in the hands of just a few mutual funds, banks, and corporations. This is the first clear picture of the global concentration of financial power, and point out the worldwide financial system's vulnerability as it stood on the brink of the current economic crisis.*

*A pair of physicists at the Swiss Federal Institute of Technology in Zurich did a physics-based analysis of the world economy as it looked in early 2007. Stefano Battiston and James Glattfelder extracted the information from the tangled yarn that links 24,877 stocks and 106,141 shareholding entities in 48 countries, revealing what they called the "backbone" of each country's financial market. These backbones represented the owners of 80 percent of a country's market capital, yet consisted of remarkably few shareholders.*

*"You start off with these huge national networks that are really big, quite dense," Glattfelder said. "From that you're able to ... unveil the important structure in this original big network. You then realize most of the network isn't at all important." The most pared-down backbones exist in Anglo-Saxon countries, including the U.S., Australia, and the U.K. Paradoxically; these same countries are considered by economists to have the most widely-held stocks in the world, with ownership of companies tending to be spread out among many investors. But while each American company may link to many owners, Glattfelder and Battiston's analysis found that the owners varied little from stock to stock, meaning that comparatively few hands are holding the reins of the entire market.*

*"If you would look at this locally, it's always distributed," Glattfelder said. "If you then look at who is at the end of these links, you find that it's the same guys, [which] is not something you'd expect from the local view."*

*Matthew Jackson, an economist from Stanford University in Calif. who studies social and economic networks, said that Glattfelder and Battiston's approach could be used to answer more pointed questions about corporate control and how companies interact.*

*"It's clear, looking at financial contagion and recent crises, that understanding interrelations between companies and holdings is very important in the future," he*

---

[19] The article can be found online at: http://www.sciencenews.org/view/generic/id/40886/title/Networks_reveal_concentrated_ownership_of_corporations.

*said. "Certainly people have some understanding of how large some of these financial institutions in the world are, there's some feeling of how intertwined they are, but there's a big difference between having an impression and actually having ... more explicit numbers to put behind it."*

*Based on their analysis, Glattfelder and Battiston identified the ten investment entities who are "big fish" in the most countries. The biggest fish was the Capital Group Companies, with major stakes in 36 of the 48 countries studied. In identifying these major players, the physicists accounted for secondary ownership—owning stock in companies who then owned stock in another company—in an attempt to quantify the potential control a given agent might have in a market.*

*The results raise questions of where and when a company could choose to exert this influence, but Glattfelder and Battiston are reluctant to speculate.*

*"In this kind of science, complex systems, you're not aiming at making predictions [like] ... where the tennis ball will be at given place in given time," Battiston said. "What you're trying to estimate is ... the potential influence that [an investor] has."*

*Glattfelder added that the internationalism of these powerful companies makes it difficult to gauge their economic influence. "[With] new company structures which are so big and spanning the globe, it's hard to see what they're up to and what they're doing," he said. Large, sparse networks dominated by a few major companies could also be more vulnerable, he said. "In network speak, if those nodes fail, that has a big effect on the network."*

*The results will be published in an upcoming issue of the journal Physical Review E.*

By Lauren Schenkman, Inside Science News Service, 26 August 2009.[20]

[20] Online at: http://www.livescience.com/culture/090826-stock-market.html.

# Curriculum Vitae

James B. Glattfelder

## Personal Information

| | |
|---|---|
| Date of Birth: | 19.08.1972 |
| Place of Birth: | Samedan, GR |
| Citizenship: | Swiss/British |

## Education

| | |
|---|---|
| 2006–2010 | PhD-student at the Chair of "Systems Design", Department of Management Technology and Economics, ETH Zurich 50 % employment (see Work Experience for remaining employment) |
| 1992–1999 | M.Sc. in (theoretical high energy particle) physics, ETH Zurich Diploma thesis: "Supersymmetry and the Higgs Sector of the Standard Model" |
| 1985–1992 | Swiss Matura in economics, Lyceum Alpinum Zuoz |

## Work Experience

| | |
|---|---|
| 2002–present | Senior researcher at Olsen Ltd (investment management industry, Zurich): R&D of automated real-time trading model algorithms for the FX market, focus on identifying statistical properties and patterns<br>From 2006 to 2010 50 % employment next to dissertation |

J. B. Glattfelder, *Decoding Complexity*, Springer Theses,
DOI: 10.1007/978-3-642-33424-5, © Springer-Verlag Berlin Heidelberg 2013

## Further Activities

| | |
|---|---|
| 2000–2001 | Trip around the world |
| since 2005 | Treasurer of the association noon.ch (charitable organization) |

## Skills

| | |
|---|---|
| Languages: | German (native), English (excellent) |
| IT: | Platforms: UNIX (Linux), Mac OS, Windows Programming languages: Java, C++, MySQL, Bash, PHP, HTML, CSS<br>Tools: Eclipse IDE, Subversion, Matlab, R, LaTeX, MS Office, Typo3 |

## Sports and Hobbies

Snowboarding, rock climbing, surfing, travel
Technology, philosophy, environmental issues, society, subculture, electronica

## Publications

### *Peer Reviewed*

Glattfelder, J. B. and Battiston, S. (2009), **Backbone of Complex Networks of Corporations: The Flow of Control**, *Physical Review E*, 80(3), 036104
Glattfelder, J. B., Dupuis, A. and Olsen, R. B. (2011), **Patterns in High-Frequency FX Data: Discovery of 12 Empirical Scaling Laws**, *Quantitative Finance*, 4(11), 599–614
Vitali, S., Glattfelder, J. B., Battiston, S. (2011), **The Network of Global Corporate Control**, *PLoS ONE*, 6(10), e25995

### *Articles*

Glattfelder, J. B., **Decoding Complexity: The Organizing Principles Behind our Economy**, *The Montréal Review*, April 2012, http://www.themontrealreview.com/2009/The-Organizing-Principles-Behind-Economy-Complex-Systems.php

## *Book Chapter*

Chapter 7:
**The Structure of Financial Networks**, Battiston, S., Glattfelder, J. B., Garlaschelli, D., Lillo, F. and Caldarelli, G.
In:
**Network Science: Complexity in Nature and Technology**, Estrada, E., Fox, M., Higham, D. J. and Oppo, G.-L. (eds.), Springer, 2010

## *Workingpapers*

The novel results given in Chap. 2 ($\bar{v}^{net}$ and $\bar{v}^{int}$)
The unpublished results given in Chap. 5

## Talks

**Global Ownership: Unveiling the Structures of Real-World Complex Networks**, Section Dynamics and Statistical Physics (DY), February meeting of the German Physical Society (DPG), Berlin, Germany, 25th February 2008

**The Backbone of Control in G8 Countries**, Annual Conference Physics of Socio-Economic Systems (AKSOE), February meeting of the German Physical Society (DPG), Berlin, Germany, 27th February 2008

**The Network of Global Corporate Control**, CCSS International Workshop, Zurich, Switzerland, 9th of June 2009

**The Power of Few: The Organizing Principles Behind our Markets**, public lecture, Treffpunkt Science City, Zurich, Switzerland, 20th of November 2011

**How Stable is our Economy? Systemic Risk and Economic Networks**, invited talk, Financial Risk Management Event, Deloitte, Zurich, Switzerland, 11th of January 2012

## Media Coverage

The two publications relating to the ownership networks attracted a lot of interest in the international media and were widely covered. See Appendix F. and http://www.sg.ethz.ch/response/index for additional information.

# References


N.A. Alves, Unveiling community structures in weighted networks. Phys. Rev. E **76**(3), 36101 (2007)

R. Albert, A.L. Barabási, Statistical mechanics of complex networks. Rev. Mod. Phys. **74**(1), 47–97 (2002)

R. Albert, H. Jeong, A.L. Barabási, Internet: diameter of the world-wide web. Nature **401**(6749), 130–131 (1999)

L.A.N. Amaral, S.V. Buldyrev, S. Havlin, M.A. Salinger, H.E. Stanley, Power law scaling for a system of interacting units with complex internal structure. Phys. Rev. Lett. **80**(7), 1385–1388 (1998)

P.W. Anderson, More is different. Science **177**(4047), 393–396 (1972)

D. Ariely, *Predictably Irrational: The Hidden Forces that Shape our Decisions* (Harper, New York, 2008)

S. Baldone, F. Brioschi, S. Paleari. Ownership measures among firms connected by cross-shareholdings and a further analogy with input-output theory, in *4th JAFEE International Conference on Investment and Derivatives* (1998)

A.L. Barabási, R. Albert, Emergence of scaling in random networks. Science **286**, 509 (1999)

O.E. Barndorff-Nielsen, K. Prause, Apparent scaling. Finance Stochast. **5**(1), 103–113 (2001)

A. Barrat, M. Barthelemy, R. Pastor-Satorras, A. Vespignani, The architecture of complex weighted networks. Proc. Natl. Acad. Sci. USA **101**, 3747 (2004)

V.D. Blondel, J.L. Guillaume, R. Lambiotte, E.L.J.S. Mech, Fast unfolding of communities in large networks. J. Stat. Mech. **2008**, P10008 (2008)

M. Botvinick, J. Cohen et al., Rubber hands ‘feel’ touch that eyes see. Nature **391**(6669), 756–756 (1998)

J.P. Bouchaud, Power laws in economics and finance: some ideas from physics. Quant. Finance **1**(1), 105–112 (2001)

S. Brin, L. Page, The anatomy of a large-scale hypertextual web search engine. Comput. Netw. ISDN Syst. **30**(1–7), 107–117 (1998)

F. Brioschi, L. Buzzacchi, M.G. Colombo, Risk capital financing and the separation of ownership and control in business groups. J. Bank. Finance **13**(1), 747–772 (1989)

G. Brun, D. Kuenzle, Lecture notes *Introduction to the Philosophy of Science*, (ETH, Zurich, 2008), http://www.vvz.ethz.ch/Vorlesungsverzeichnis/lerneinheitPre.do?lerneinh

M. Buchanan, G. Caldarelli, A networked world. Phys. World **23**, 22–24 (2010)

G. Caldarelli, *Scale-Free Networks: Complex Webs in Nature and Technology* (Oxford University Press, London, 2007)

G. Caldarelli, A. Capocci, P. de Los Rios, M.A. Muñoz, Scale-free networks from varying vertex intrinsic fitness. Phys. Rev. Lett. **89**(22), 258702 (2002)

A. Capocci, V.D.P. Servedio, G. Caldarelli, F. Colaiori, Detecting communities in large networks. Physica A **352**(2–4), 669–676 (2005)

J.L. Casti, *Alternate Realities: Mathematical Models of Nature and Man* (Wiley-Interscience, New York, 1989)

A. Clauset, M.E.J. Newman, C. Moore, Finding community structure in very large networks. Phys. Rev. E **70**(6), 66111 (2004)

L.F. Costa, F.A. Rodrigues, G. Travieso, P.R.V. Boas, Characterization of complex networks: a survey of measurements. Adv. Phys. **56**(1), 167–242 (2007)

M.J. Crockett, L. Clark, G. Tabibnia, M.D. Lieberman, T.W. Robbins, Serotonin modulates behavioral reactions to unfairness. Science **320**(5884), 1739–1739 (2008)

M.M. Dacorogna, R. Gencay, *An Introduction to High-Frequency Finance* (Academic Press, San Diego, 2001)

D. Deutsch, *Beginning of Infinity: Explanations That Transform the World* (Viking, New York, 2011)

L. Donetti, M.A. Munoz, Detecting network communities: a new systematic and efficient algorithm. J. Stat. Mech. **2004**, P10012 (2004)
S.N. Dorogovtsev, J.F.F. Mendes, Evolution of networks. Adv. Phys. **51**(4), 1079–1187 (2002)
S.N. Dorogovtsev, J.F.F. Mendes, *Evolution of Networks: From Biological Nets to the Internet and WWW* (Oxford University Press, Oxford, 2003)
S.N. Dorogovtsev, J.F.F. Mendes, A.N. Samukhin, Giant strongly connected component of directed networks. Phys. Rev. E **64**(2), 25101 (2001)
T. Di Matteo, Multi-scaling in finance. Quant. Finance **7**(1), 21–36 (2007)
D. Eagleman, *Incognito: The Secret Lives of the Brain* (Phanteon, New York, 2011)
J.D. Farmer, F. Lillo, On the origin of power-law tails in price fluctuations. Quant. Finance **4**(1), 7–11 (2004)
P.K. Feyerabend, *Against Method: Outline of an Anarchistic Theory of Knowledge* (NLB, London, 1975)
X. Gabaix, P. Gopikrishnan, V. Plerou, H.E. Stanley, A theory of power-law distributions in financial market fluctuations. Nature **423**(6937), 267–270 (2003)
S. Galluccio, G. Caldarelli, M. Marsili, Y.C. Zhang, Scaling in currency exchange. Physica A **245**(3–4), 423–436 (1997)
D. Garlaschelli, G. Caldarelli, L. Pietronero, Universal scaling relations in food webs. Nature **423**(6936), 165–168 (2003)
M.M. Geipel. *Dynamics of Communities and Code in Open Source Software*. Ph.D. Thesis, Chair of Systems Design, ETH, Zurich, 2010. http://e-collection.ethbib.ethz.ch/view/eth:254?q=mmgeipel
D.T. Gilbert, *Stumbling on Happiness* (Vintage Books, New York, 2006)
J.B. Glattfelder, A. Dupuis, R.B. Olsen, Patterns in high-frequency FX data: discovery of 12 empirical scaling laws. Quant. Finance **11**(4), 599–614 (2011)
K. Gödel, Über formal unentscheidbare Sätze der Principia Mathematica und verwandter Systeme. Monatshefte für Mathematik und Physik **38**, 173–198 (1931)
D.M. Guillaume, M.M. Dacorogna, R.R. Davé, U.A. Müller, R.B. Olsen, O.V. Pictet, From the bird's eye to the microscope: a survey of new stylized facts of the intra-daily foreign exchange markets. Finance and stochast. **1**(2), 95–129 (1997)
C. Hillman, *A Categorical Primer* (University of Washington, Seattle, 2001)
R.A. Horn, C.R. Johnson *Matrix Analysis* (Cambridge University Press, Cambridge, 1990)
D. Kahneman, *Thinking* (Farrar, New York, 2011)
M. Kosfeld, M. Heinrichs, P.J. Zak, U. Fischbacher, E. Fehr, Oxytocin increases trust in humans. Nature **435**(7042), 673–676 (2005)
M. König, S. Battiston. From graph theory to models of economic networks: a tutorial, *Networks, Topology and Dynamics*, ed. by A.K. Naimzada, S. Stefani, A. Torriero (Springer, Berlin 2009)
T.S. Kuhn, *The Structure of Scientific Revolutions* (University of Chicago Press, Chicago, 1962)
H. Kwak, C. Lee, H. Park, S. Moon, What is twitter, a social network or a news media? WWW, 2010. http://an.kaist.ac.kr/traces/WWW2010.html
D. Leech, Corporate ownership and control: a new look at the evidence of berle and means. Oxf. Econ. Pap. **39**(3), 534–551 (1987)
G.W. Leibniz, *Principes de la Nature et de la Grâce fondés en raison (1714)* (Presses Universitaires de France, Paris, 1954)
S. Lloyd, *Programming the Universe* (Knopf, New York, 2006)
T. Lux, Financial power laws: empirical evidence, models, and mechanisms. Power Laws Soc. Sci. (2005)
X. Ma, S. Zotter, J. Kofler, R. Ursin, T. Jennewein, Č. Brukner, A. Zeilinger, Experimental delayed-choice entanglement swapping. Nat. Phys. **8**(6), 480–485 (2012)
B. Mandelbrot, The variation of certain speculative prices. J. Bus. **36**(4), 394–419 (1963)
R.N. Mantegna, H.E. Stanley, Scaling behaviour in the dynamics of an economic index. Nature **376**(6535), 46–49 (1995)

T. Metzinger, *The Ego Tunnel: The Science of the Mind and the Myth of the Self Basic Books*, (Springer, Berlin, 2010)
S. Milgram, The small world problem. Psychol. Today **2**(1), 60–67 (1967)
U.A. Müller, M.M. Dacorogna, R.B. Olsen, O.V. Pictet, M. Schwarz, C. Morgenegg, Statistical study of foreign exchange rates: empirical evidence of a price change scaling law, and intraday analysis. J. Bank. Finance **14**(6), 1189–1208 (1990)
M.E.J. Newman, The structure and function of complex networks. SIAM Rev. **45**(2), 167–256 (2003)
M.E.J. Newman, Detecting community structure in networks. Eur. Phys. J. B **38**(2), 321–330 (2004)
M.E.J. Newman, Power laws, Pareto distributions and Zipf's law. Contemp. Phys. **46**(5), 323–351 (2005)
M.E.J. Newman, Finding community structure in networks using the eigenvectors of matrices. Phys. Rev. E **74**(3), 36104 (2006)
M.E.J. Newman, D.J. Watts, Scaling and percolation in the small-world network model. Phys. Rev. E **60**(6), 7332–7342 (1999)
M.E.J. Newman, S.H. Strogatz, D.J. Watts, Random graphs with arbitrary degree distributions and their applications. Phys. Rev. E **64**(2), 26118 (2001)
M.E.J. Newman, D.J. Watts, S.H. Strogatz, Random graph models of social networks. Proc. Natl. Acad. Sci. USA **99**(90001), 2566–2572 (2002)
M.E.J. Newman, A.L. Barabási, D.J. Watts, *The Structure and Dynamics of Networks* (Princeton University Press, Princeton, 2006)
V. Pareto, The new theories of economics. J. Polit. Econ. **5**(4), 485–502 (1897)
J. Park, M.E.J. Newman, Statistical mechanics of networks. Phys. Rev. E **70**(6), 66117 (2004)
R. Pastor-Satorras, A. Vázquez, A. Vespignani, Dynamical and correlation properties of the internet. Phys. Rev. Lett. **87**(25), 258701 (2001)
V.I. Petkova, H.H. Ehrsson, If i were you: perceptual illusion of body swapping. PLoS One **3**(12), e3832 (2008)
R.D. Pirsig, *Zen and the Art of Motorcycle Maintenance: An Inquiry into Values* (William Morrow, New York, 1974)
M. Planck, *Scientific Autobiography and Other Papers* (Williams& Norgate, London, 1950)
H. Plassmann, J. O'Doherty, B. Shiv, A. Rangel, Marketing actions can modulate neural representations of experienced pleasantness. Proc. Natl. Acad. Sci. **105**(3), 2008 (1050)
K.R. Popper, *Realism and the Aim of Science*, vol. 1, (Rowman and Littlefield, Totowa, 1983)
R.D. Precht, *Wer bin ich - und wenn ja, wie viele?* (Goldmann, Munchen, 2007)
J. Reichardt, S. Bornholdt, Statistical mechanics of community detection. Phys. Rev. E **74**(1), 16110 (2006)
P.E.A. Rényi, P. Erdős, On the evolution of random graphs. Publ. Math. Inst. Hung. Acad. Sci. **5**, 17–61 (1960)
J.A. Schülein, S. Reitze, Wissenschaftstheorie für Einsteiger. WUV UniversitStsverlag, 2002
E. Seneta, *Non-Negative Matrices and Markov Chains* (Springer, New York, 2006)
W. Singer, *Der Beobachter im Gehirn* (Suhrkamp, Frankfurt am Main, 2002)
A.D. Sokal, Transgressing the boundaries: toward a transformative hermeneutics of quantum gravity. Soc. Text **46**, 217–252 (1996)
D. Sornette, *Critical Phenomena in Natural Sciences* (Springer, New York, 2000)
D. Sornette, Interdisciplinarity in socio-economics, mathematical analysis and predictability of complex systems. Arxiv preprint arXiv:0807.3814, 2008. http://arxiv.org/abs/0807.3814v1
A. Stefanov, H. Zbinden, N. Gisin, A. Suarez, Quantum entanglement with acousto-optic modulators: two-photon beats and bell experiments with moving beam splitters. Phys. Rev. A **67**(4), 042115 (2003)
S.H. Strogatz, *Non Linear Dynamics* (Addison Wesley, Reading, 1994)
S.H. Strogatz, Exploring complex networks. Nature **410**(6825), 268–276 (2001)

R. Tarnas, *The Passion of the Western Mind: Understanding the Ideas that Have Shaped Our World View* (Ballantine Books, New York, 1993)

U.C. Täuber. Field-theoretic methods. Arxiv preprint arXiv:0707.0794v1, 2008. http://arxiv.org/abs/0707.0794

D.J. Watts, S.H. Strogatz, Collective dynamics of 'small-world' networks. Nature **393**(440), 440–442 (1998)

S. Weinberg, Steven Weinberg on scientific revolutions. NY Rev. Books **45**(15), (1998)

D.B. West, *Introduction to Graph Theory* (Prentice Hall, New York, 2001)

G.B. West, J.H. Brown, B.J. Enquist, A general model for the origin of allometric scaling laws in biology. Science **276**(5309), 122 (1997)

J.A. Wheele, Information, Physics, Quantum: The Search for Links. *Complexity, Entropy, and the Physics of Information*, ed. by W. Zurek, (Addison Wesley, Reading, 1990)

E. Wigner, The unreasonable effectiveness of mathematics in the natural sciences. Commun. Pure Appl. Math. **13**(1), 222–237 (1960)

S. Wolfram, *A New Kind of Science* (Wolfram Media Inc, Champaign, 2002)

A. Zeilinger, *Einsteins Schleier Die neue Welt der Quantenphysik* (C.H Beck, München, 2003)

The manufacturer's authorised representative in the EU is Springer Nature Customer Service Centre GmbH, Europaplatz 3, 69115 Heidelberg, Germany. If you have any concerns regarding our products, please contact ProductSafety@springernature.com

Printed and bound by CPI Group (UK) Ltd, Croydon, CR0 4YY
22/09/2026
02221062-0008